DESTINATION MOON

Pallava Bagla is Science Editor, NDTV (New Delhi Television) and Chief Correspondent (South Asia) for the American weekly *Science*. In 2005 he was honoured with the National Award for best science journalism in the print media by the Indian government and in 2003 he was awarded for 'outstanding journalism' by the United Nations. He is also a skilled photographer and images taken by him have appeared in *National Geographic*, *Time*, *Newsweek*, *Nature* and *New Scientist*. He has covered the Indian space and nuclear programme for the past two decades.
pallava.bagla@gmail.com

Dr Subhadra Menon is an award-winning science and health writer and a former correspondent for *India Today*. She has also contributed to *New Scientist* and *Scientific American*. She is the author of *No Place to Go*, a book on India's ailing health sector. She currently works with the Public Health Foundation of India, spearheading its communication efforts.
subhadra.menon@gmail.com

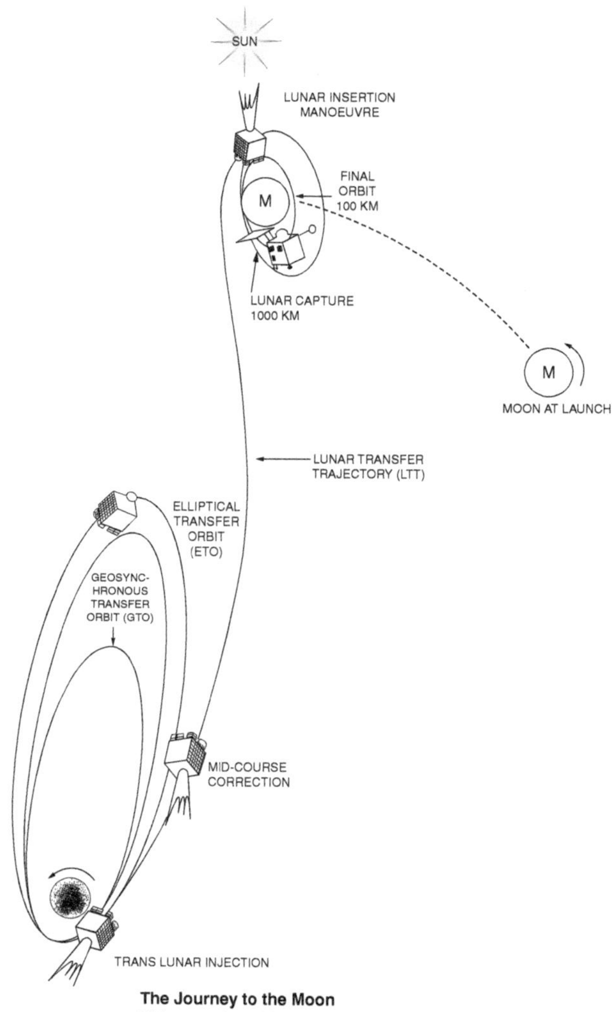

The Journey to the Moon

This route is subject to change closer to the mission date.

Courtesy : ISRO

DESTINATION MOON

India's Quest for the Moon, Mars and Beyond

Pallava Bagla
&
Subhadra Menon

HarperCollins *Publishers* India

First published in India by
HarperCollins *Publishers* in 2008
A-75, Sector 57, Noida, Uttar Pradesh 201301, India
www.harpercollins.co.in

1 2 3 4 5 6 7 8 9 10

P-ISBN: 978-81-7223-676-2
E-ISBN: 978-93-5116-064-9

Typeset in 11/14 Minion
Mindway Design

To Karan Pathak, close friend and son of an Air Force officer. With Karan we took many early flights of fancy that explored the magic of science, but he left us suddenly when his life was cut short in the Patna air crash of 2000

≋

To Shantnu Sharma, also close friend, sailor and explorer, who taught us that life is one adventurous journey—for always being there

Contents

FOREWORD
Space Odyssey

Humankind's odyssey into the final frontier, space, began half century ago. It was on 4 October 2007 that the world observed the fiftieth anniversary of the launch of Sputnik, the first artificial satellite to orbit the earth. These fifty years of space research have influenced a transformation of humankind in a manner that is unprecedented in respect of its scientific, technological and exploratory outreach. To cite only a few related examples: near-earth orbiting satellites have been used for the scientific exploration of the earth and its environment, for solar system studies and the observation of the distant regions of the universe. Satellite-borne imaging systems today routinely observe the earth for enabling timely, accurate and precise information about natural resources. Further, geosynchronous satellites extensively carry out communications, broadcasting as well as weather observations.

Complex scientific missions have been undertaken to explore different planets and their moons in our solar system and to understand and answer fundamental questions related to their origin and evolution. In fact, the first such spacecraft has already crossed the outer fringes of the solar system, the boundary of the heliosphere where the effects of solar radiation are overtaken by those of the interstellar medium. Moreover, humankind has also demonstrated the ability to live in outer space for long durations through the establishment of manned space stations, including the most recent International Space Station. Man has set foot for the first time on another object in the solar system, the nearby moon. All of these accomplishments put together have shaped an important dimension of the history of evolution of space and the ability to access space, both through manned and unmanned vehicles. In a nutshell, the first five decades of the global space agenda have been full of exciting discoveries and technological feats, the likes of which have been rarely witnessed in any other domain of human endeavour.

Against this backdrop, the exploration of planets and other solar system objects has been one of the most rewarding scientific experiences. Many discoveries have been made, raising new scientific questions and throwing up interesting technological challenges, thereby providing fresh impetus to the exploration of planetary systems. One crucial scientific issue is the presence of water on many planetary bodies – around the poles of Mercury, possibly also the poles of the moon and Mars, as well as several satellites of Jupiter such as Europa, Ganymede and Callisto. The discovery of primitive objects beyond the orbit of Neptune in the solar system is another exciting development. Astronomical evidence of protoplanetary discs depicting bipolar flow from the central star and x-wind coming from the sun during the early stages of the solar nebula is yet another interesting discovery.

The earth's geological history holds evidence that severe mass extinctions were caused due to the impact of asteroids, comets and/or other near-earth objects. This evidence needs further investigation, not only because of its inherent scientific value but also because a similar impact from a large object in the future could wipe out civilization in one stroke. The presence of amino acids and such other large organic molecules, considered to be the building blocks of life, in some meteorites is closely related to the issue of the origins of life outside earth.

India's own space programme started in the early sixties with the scientific objectives of exploring the earth's atmosphere and ionosphere, and studying astrophysical objects using rockets. The next four decades saw the programme steadily evolve into a full-fledged multi-dimensional effort encompassing satellites, launch vehicles, space science and applications. Today India has developed its own autonomous access to space through its two launch vehicles, the Polar Satellite Launch Vehicle (PSLV) and Geosynchronous Satellite Launch Vehicle (GSLV). PSLV is primarily used for orbiting Indian Remote Sensing satellites of 1 to 1.5-ton class, into polar sun-synchronous orbits. The Remote Sensing Satellites (IRS) designed and developed in India form one of the largest and most sophisticated constellations in the civilian domain for earth resource survey applications. Similarly, present versions of the GSLV can place the 2-2.5-ton class of the multi-purpose INSAT into geosynchronous transfer orbits. INSATs are used for communication, providing connectivity to different parts of the country, and rendering broadcasting services towards entertainment, education and health, in addition to making meteorological observations for predicting weather phenomena. These are some of India's satellite and launch vehicle capabilities that provide scientists with a unique opportunity to undertake ambitious space science

experiments in astronomy and astrophysics, besides studies of the earth and its environment as well as planetary science.

Coming to planetary exploration, it is of interest to recall that from the early 1960s, several groups in India have been carrying out research in planetary sciences using telescopes and laboratory simulations. Further, the study of cosmic ray effects on meteorites was pioneered by Prof. Devendra Lal and Prof. Narendra Bhandari initially at the Tata Institute of Fundamental Research (TIFR), Mumbai and later at the Physical Research Laboratory, Ahmedabad. With the arrival of lunar samples from Apollo and Luna missions, the early investigations at TIFR were extended to rocks and dust from the moon during 1972–1981.

It is against this backdrop of diverse early scientific endeavours and contemporary technological capabilities that the idea of India undertaking a mission to the moon was mooted. The occasion was a lecture that I gave during the first Technology Day at Delhi on 11 May 1999, held to celebrate the first anniversary of Pokharan II. This was organized at the suggestion of Prof. Murli Manohar Joshi, the minister for human resource development and science and technology. The meeting was attended by a wide cross-section of Delhi's intelligentsia including academics, public servants and eminent citizens as well as ministers and members of parliament, and the media. My presentation mostly concentrated on the latest developments in India's space programme and certain aspects of our future plans. I had included in this material a specific brief on the capabilities of PSLV and GSLV in the context of undertaking planetary exploration missions. I mentioned that the PSLV can place a spacecraft of 150 kg in a 1000 km orbit around Mars and that GSLV can place about 300 kg in such an orbit. A similar assessment for Venus would be 110 kg and 220 kg respectively, for these two vehicles. In the context of the moon, the possibility of orbiting a 500-kg spacecraft with the PSLV capability was pointed out. The purpose of mentioning the capabilities of PSLV and GSLV as well as the maturity of the satellite technology was to highlight that India was in a position to undertake a meaningful planetary mission. It caught the imagination of the people who attended the lecture and also the media. I vividly recall Pallava Bagla, one of the authors of this book, prominently featuring those details on the front page in the next day's *Indian Express*. Not only did he give an excellent account of the lecture but also elaborated aspects of the planetary mission and even made the suggestion that India undertake a lunar mission. Encouraged by the response of the public in general to this possibility, we decided that we should go through a process of consultation with the scientific community for assessing the usefulness of the lunar mission in the context of India's space exploration initiatives.

Accordingly, a discussion on Indian plans for a mission to the moon was organized at the annual meeting of the Indian Academy of Sciences at Lucknow in November 1999. Among other things the fellows of the Academy were presented specific details of the planned mission objectives, in the context of the achievements of similar missions in the past. Existing, extensive information on the physical, chemical, geological and geochronological aspects of the moon from past missions were assessed so that Indian experiments could focus on areas where gaps in our knowledge exist. The Indian scientific community was also briefed on the planned use of contemporary instruments for observations as well as spacecraft configuration in addition to the overall mission details. The symposium also highlighted international ramifications as well as contemporary issues in legal and international cooperation domains. There was wholehearted support from the scientific community to the idea of this mission, considering its scientific worthiness, technological challenges, inspirational potential as well as its strategic dimension.

As the next step, the Astronautical Society of India in its annual meeting in Ahmedabad on 4 February 2000 was presented with the plans for the lunar mission. The specific challenges that such a mission could pose in developing improved control, guidance and navigation capabilities, trajectory analysis and realizing higher levels of sophistication in celestial mechanics, as well as new areas of technologies for deep space communications, were recognized by the technological community as strong motivating factors.

In order to sensitize the political spectrum about such an important national initiative, a brief outline of ISRO's ideas in the area of planetary exploration and the possibility of a lunar mission in particular was presented to the Standing Committee of the Parliament on Science and Technology, Environment and Forests, on 9 April 2002. This committee, in its hundredth report, presented to the Rajya Sabha and the Lok Sabha on 2 May 2002, observed that space science and environmental research have been important component of the Indian space programme. It further stated that it was guided by the consideration of the Department of Space that India should obtain technological superiority and should have the capability of sending a craft to the moon. It observed that the proposed mission to the moon was expected to provide an impetus to the advancement of science in India and a challenge to technologists. Further, it could also serve as a test-bed for future missions that India could undertake to explore the outer world in the new millennium, thus providing challenging opportunities to the younger generation of scientists. The committee recognized that the moon was going to be the target of international attention in the coming decades as a takeoff point for planetary explorations. The committee was therefore of the view that India should possess the capability of reaching the moon so that it could assert its right and presence there. The committee went on to recommend that India should go step by step from orbiters to landers, and then to crafts which could return with samples, before it could think of a manned mission.

The approach as described above enabled an assessment of interest – from scientists, technologists and also the political system – and led us to the conclusion that it was appropriate and timely to conduct a detailed study of all the elements of such a mission. A National Task Force was accordingly constituted to examine related issues. This taskforce, under the chairmanship of Dr George Joseph, former director of ISRO's Space Application Centre, consisted of experts drawn from different disciplines relevant to the study. As a result, a report covering scientific, technical and human resources, as well as the details of the mission, budget and schedule were all assessed, and finally led to the conclusion that a lunar mission could be undertaken with a high level of confidence by ISRO. This conclusion was based specifically on the spacecraft and launch vehicles heritage of ISRO and also the level of maturity of several types of instruments that could be flown on such a mission.

The report submitted by the National Task Force was finally reviewed by a broad-based national committee consisting of scientists, technologists and policymakers, headed by Prof. M.G.K. Menon. This committee, after due deliberation, came to the conclusion that ISRO had the necessary expertise to develop and launch the moon mission with innovative features, including new instruments, that would be different from past missions. The committee felt that apart from technological challenges and scientific opportunities in the area of planetary research, the project could provide new avenues to attract young talent to the area of fundamental research. The committee suggested that ISRO should take the steps necessary to involve universities right from the beginning of the programme. Further, it would be necessary to create a cadre of scientists who would be ready to receive the vast amount of data and analyze it as the mission got into its operational phase.

The Space Commission at its meeting held on 6 June 2003 deliberated on the recommendation of the Space Department to undertake a mission to the moon based on viable scientific objectives, providing at the same time a long-term perspective for India's ambitions to be part of a planetary exploration endeavour of the global community. Acknowledging the efforts that had gone through several preparatory steps as mentioned in the foregoing paragraphs, the commission opined that it was timely and appropriate to embark on a new initiative relating to planetary exploration and towards this, the suggested lunar mission was the most logical first step.

On the occasion of Independence Day, it is customary for the prime minister to make a few announcements regarding important future plans of the government in his address to the nation from the ramparts of the Red Fort in Delhi. Considering the importance of India's planned initiatives in planetary exploration and its determination to play a key role in the planetary exploration endeavour of humankind in the twenty-first century, Prime Minister Atal Behari Vajpayee in his address to the nation on 15 August 2003 announced the country's intention to plan for a mission to the moon. Even though ISRO had characterized this mission only as Chandrayaan, the prime minister thoughtfully converted it to Chandrayaan-1. He clarified to me later, that it could not be a single-shot effort but needed to have a long-term perspective. Thus was born India's maiden mission to the moon, Chandrayaan-1, after nearly four years of preparatory work.

Chandrayaan-1 is primarily designed to provide unique, high quality global data on the chemistry and mineralogy of the moon. A novel concept to provide a slot for international experiments has yielded excellent results in attracting additional payloads on Chandrayaan-l that complement the Indian experiments very well. This suite of eleven experiments on Chandrayaan-l covers nearly the full range of scientific investigations, including detailed topography and ranging that will yield the best digital elevation map of the moon, global mapping of elemental abundance and mineral concentrations from spectroscopic studies, the search for water ice at the poles, neutral atom mapping, cislunar radiation environment studies, mass spectrometry very close to the lunar surface and direct measurements of residual radioactivity that provide evidence for volatile transport on the lunar surface.

This mission provides the critical impetus for Indian scientists to further explore the moon as well as other objects in our solar system. All these exploratory programmes come under the overall research plan 'for understanding the origin and evolution of solar system objects'.

Beyond Chandrayaan-1, lunar exploration strategies may have to address in situ observations that examine the regolith and bedrock through deep drilling or trenching techniques, providing a more detailed view of the evolution of the lunar surface. These are currently being examined by various future programmes under consideration by many nations. Beyond the moon, Mercury and Mars appear to be ideal planets to pursue detailed studies in the post-Messenger and Beppi-Colombo era. In the outer solar system, the Jovian satellites Europa and Ganymede appear to be prime targets for future exploration internationally.

The Indian planetary exploration programme beyond Chandrayaan-1 includes a near-term mission Chandrayaan-2 to address in situ studies as well as to improve upon global data from an orbiter platform. A Mars orbiter around 2015, with the primary goal of understanding Martian atmospheric processes and weather/dust storms, the Martian ionosphere, the effect of solar wind, surface magnetic fields, and a search for paleo-water and surface resources is under consideration. An asteroid orbiter and comet flyby is conceived for 2017, to study the evolution of asteroids and comets, early solar system processes, the meteorite-asteroid connection, and the physical and chemical properties of asteroid and cometary material.

As we move towards undertaking increasingly complex and diverse approaches for planetary exploration, there could obviously be questions of resources, both financial and human.

It is in this context that India has to develop unique strategies of cooperation and collaboration with other nations with similar interests. Through such partnerships India could widen its role in the human endeavour of solar system exploration in the twenty-first century on an equal partnership basis. Similarly, cost-effective missions like Chandrayaan-1 could create the necessary credentials, both in terms of technology capability and the ability to play a unique scientific role, that in turn will enable India to play its rightful role in the Comity of Nations in the most exciting area of human exploration.

Dr Krishnaswamy Kasturirangan
Member of Parliament (Rajya Sabha)
and Former Chairman, ISRO

July 2008
New Delhi

Foreword
ISRO's Big Leap

The moon has been an object of fascination from time immemorial for poets, romantics and the common man. For scientists, the moon is an object of intense interest because of the mystery associated with its role in the evolution of our own planet, of which it is the only natural satellite. The moon aroused keen interest in our ancient sages and one can find a mention of it in almost all religious mythologies. The Italian astronomer Galileo Galilei was the first to observe the moon through a telescope and reveal the nature of its surface.

The moon is a part of our lives, in more ways than one. When I was a child of about three, we lived in a small village called Thirunanthikara, now in Kanyakumari district. I have distinct memories of sitting out under the open sky on my grandmother's lap, with her trying to feed me. She would often promise to get me the moon in return for my being a 'nalla kutti' (good boy), and polishing off everything on my plate. The clear full moon in that unpolluted atmosphere always made me wonder why there were so many strange shapes on its surface. It was as though you could see whatever your imagination conjured up, so if one day it was a deer walking across, another day it could be a man sitting there. My love for the moon and the desire to unravel the many unknowns about outer space have been gathering pace since then.

The race for lunar exploration received tremendous impetus during the 1960s, with US President John F. Kennedy giving a mandate to NASA to put a man on the moon by the end-1960s. The Apollo programme, with vast resources and the availability of the Saturn V rocket, culminated in the first human landing on the moon on 21 July 1969. The erstwhile Soviet Union also undertook several pioneering and successful unmanned missions to study the moon. Following numerous attempts by the two major space powers of the Cold War era to study the moon, there was a lull in this regard for about two decades. With the dawn of a new renaissance in the 1990s pertaining to lunar exploration, several countries, especially relatively new entrants such as the European Space Agency, India, Japan and China are undertaking missions to explore the moon with renewed vigour and interest. Today, there are ongoing lunar missions such as Kaguya from Japan and Chang'e from China.

India is undertaking its first unmanned lunar mission, Chandrayaan-1, for studying several aspects of the moon including the far side of the moon, which is yet to be explored in detail. Chandrayaan-1 is bound to generate tremendous interest among academicians, researchers and scientists, and ignite the minds of young and old in the country. In addition, Chandrayaan-1 is truly international in character. While five of its payloads are from India, six payloads from the US, Europe and Bulgaria are also being flown aboard this remote sensing mission. The preparatory groundwork and justification were so thorough that final governmental decisions and approvals

for us to undertake the mission came through quickly, as is often the case with ISRO's many initiatives. The support provided by successive prime ministers, who have always held the space portfolio for the country, is also critical in achieving these landmarks.

In India, probably after the launch of the Satellite Launch Vehicle (SLV-3) and Aryabhata, it is Chandrayaan-1 which has generated tremendous enthusiasm among scientists and the general public. This is in many ways a second vision for ISRO, moving as we are from space activities oriented towards societal application to the excitement of scientific exploration of the moon, Mars and beyond. It is indeed an important phase in the Indian space programme as it provides a fresh impetus for ISRO's several thousand hard-working scientists and engineers to push the limits of their skills and expertise to challenging tasks.

There is an urgent need to understand the details of the origin of the moon, the role played by the moon in the past in the earth-moon system and its effects on the earth. It is also necessary to know about the missions to the moon undertaken by other countries, the benefit that has accrued to the scientific community from these missions and the experiments planned in recent missions to the moon. Outer space will surely be the next frontier for mankind and we have to be a part of this excitement. India has plans to continue Chandrayaan, with an unmanned lander and rover with more sophisticated instruments that would include sample analysis on the moon surface itself. Plans to undertake an unmanned mission to Mars are also unfolding. Preparations for a mission to study the sun, aptly called Aditya, are also underway and this specialized sun-observing satellite may be realized as early as 2012. These indeed are very exciting times for the Indian spacefaring community.

After fulfilling the dreams of Dr Vikram Sarabhai, the founding father of the Indian space programme, to achieve self-reliance in this critical technology area and take the fruits of space technology to the common man, ISRO is poised to enter a new phase. Chandrayaan-I is this next phase of space vision for the country, wherein ISRO is aiming to undertake planetary exploration, initiatives for a manned mission, low-cost access to space, and more sophisticated earth observation and communication satellites towards providing operational services for national development in an efficient and cost-effective manner.

On 26 May 1999, India launched a set of foreign satellites using the PSLV-C2, from Sriharikota in Andhra Pradesh. Shri Atal Behari Vajpayee, who was then prime minister of India, addressed a press conference at India's spaceport after the successful launch. The very first question asked by a journalist at the event was about India's plans for a mission to the moon, which would later be called Chandrayaan-1. Shri Vajpayee responded in his usual poetic style, saying that the moon is such a beautiful object from far, but people who have visited it have not found it so. The journalist who asked that question is one of the authors of this book. Pallava Bagla has been a prolific writer, photographer and now a veteran communicator on science and technology in general and space activities in India in particular. His efforts have not gone unnoticed. In 2006 he was honoured by the Government of India with the National Award for Outstanding Effort in Science and Technology Communication in Print Medium, the highest honour of its kind for science journalism in India awarded by the ministry of science and technology. Dr Subhadra Menon is a versatile writer who has covered a variety of topics of interest, effectively conveying these to the general public as well as to specialists, with simplicity.

This book by Pallava and Subhadra is a good step in this direction. Their lucid style of narration of complex scientific and technical principles deserves appreciation. While offering a detailed description of the mission, the book also provides information about the moon's origin and evolution and its significance in our ethos and culture. It contains well-compiled data on various facets of lunar exploration in different countries, sketches out the main aspects of the Indian space programme and also provides highlights of future missions. I compliment the authors for attempting the challenging task of writing such a book at this very important juncture. I am sure that the book will be useful for the general public and for students, and one that will serve as a good source of information for scientists too. Enjoy it and be a part of this unfolding adventure.

Dr G. Madhavan Nair | July 2008
Chairman, ISRO | Bangalore

Preface
Moonstruck

This book contains the hitherto untold story of India's journey to the moon, and may be termed a layman's guide to the Indian lunar mission. Coming as it does at a time when the country is at an important juncture in its space programme, intent on pushing the envelope on the exploratory challenges that outer space poses, it presents a ringside view of events leading up to the moon launch. The effort has been to bring to the reader a comprehensive description of Chandrayaan-1, India's maiden mission to the moon. The endeavour has also been to place the Indian moon mission in the broader context of the space programme with its dramatic early history, its slew of successes and some failures that marked the coming of age of Indian space science. In April 2008, India's space programme set a world record, sending 10 satellites into outer space using an Indian rocket launcher. If this is spectacular, there have been dramatic failures as well, such as the larger Indian rocket blowing up in July 2006 over the Bay of Bengal.

Space ventures, the science of planetary exploration and the quest to understand the moon in particular, have always made for fascinating reading. Adding to the excitement in the early decades of the third millennium will be a dash of business, with nations and people worldwide envisioning a time when outer space and planetary bodies may become repositories of resources that could support human civilization on earth. Within the broader realm, lunar exploration has always held a special place in the hearts and minds of human beings: the moon is, after all, earth's closest neighbour.

The Indian space programme has to its credit several achievements but the story-telling around many of these has remained within a rather restricted realm. Yet, the magic of reaching out into the boundlessness of outer space is undeniable and the truth is that ordinary people have a great propensity for absorbing seemingly complicated scientific facts. Moreover, subjects such as space exploration are attractive because they seem to hold out promise of an exciting future, not to mention the tangible benefit of taxpayers' money being well spent, for a change.

More importantly, science applies to all kinds of people, young and old. The spectacle at Sriharikota – the mud flats on India's eastern coast from where Indian rockets are launched into – of the crowds of ordinary people waiting to see the big bird in the sky is enough indication of the magic of space science. The air fills with that characteristic roar of rocket engines and the horizon lights up with the huge fire tail a launcher leaves behind, visible for several kilometres around the Indian space port. Sriharikota and its satellite launches are part of local legend now, with each event being viewed by hundreds of spectators.

India's dream to go to the moon is now a plan and a mission; this book is a salute to the scientific curiosity that propelled this mission. As it became clearer and clearer that India was earnest about its forthcoming lunar mission, it also became evident to us that the story of India's quest for the moon needed to be told.

Behind this book lie years of perseverance to understand one of the country's most complex national initiatives, the space programme, that conducts most of its activities behind closed doors. As a science journalist who takes seriously his job of telling it the way it is, one was allowed an insider's view over the years. From early questions critiquing the space programme to several news breaks dotting the landscape of science and technology reporting in India, the journey has not been easy, but it has been constantly rewarding. And a time came when one started to feel like a walking encyclopaedia, dying to tell a story, knitting together facts and events, plans and designs, to offer a broad brush that would be useful to all. This book has come about after more than 15 years of reporting and is backed by exhaustive reading and research, interviews with some of the movers and shakers in the programme, and a meticulous review of technical descriptions by experts.

The book has been written for young, curious minds, and also for interested adults who are fascinated by the science and exploration of space. It proceeds from a description of what we know of the origins of the moon and its evolution, through its role in the human, mainly Indian, ethos and culture to a concise description of lunar missions in other countries of the world. The first part of the book attempts to set forth the importance of the moon, and describes those events in post-Independence India that were made into milestones simply because of the way the space programme has progressed over the last four decades and more. The last three chapters focus on Chandrayaan-1 itself, and the road ahead.

It is important to state at the outset what this book is not. It is not a tome on the space programme, and makes no pretence of being an encyclopaedic compilation of the programme's many milestones, or of the scores of lunar missions conducted by other nations in the past, what they gave us and what they did not. There are already published documents, excellent websites and communication materials devoted to examining detail and presenting exhaustive information on the subject.

Neither is the book a critical analysis of success and failure. It is, rather, a celebration of Indian science and its achievements. The definition of success is itself amorphous in the context of space science. For instance, if there is a copybook launch but the satellite falls short of expectations in its optimal function, it may not be entirely accurate to term it a failure. In the same spirit, the Indian lunar mission is an exploratory venture, the real returns of which may become tangible only years from now.

Pallava Bagla

July 2008
New Delhi

Introduction

Three things cannot be long hidden: the sun, the moon, and the truth.

– Gautama Buddha

Phool Bagh, Kanpur, Uttar Pradesh. The early 1970s. A little grey rock was visiting the city and people were flocking to see it. A woman of modest financial resources who treasured learning and knowledge, a geographer teaching at the local university, decided to take her two sons to the exhibition. After all, the rock was no ordinary piece; it was a moon rock, one among the many brought down to earth by Neil Armstrong, the first man to walk on the moon. The little piece of moon rock – encased in glass with fluorescent spotlights trained on it – had pride of place in the exhibition. The rest of the hall displayed posters and pictures, with 'man on moon' and other stories. The younger of the two boys stood riveted in front of the rock. He was drawn to it for reasons even he couldn't fathom. And its impact on him would be recognized only much later.

For the very young, the moon stands for magic, mystery and all sorts of unknown things. In the mind of a child, the moon is a friendly, shiny disc that lights up the otherwise unrelenting darkness. Millions of Indian children grow up hearing stories and songs about chandamama in Hindi, or chandiran in Tamil, or ambilimaman in Malayalam. For the slightly older child, the moon becomes a symbol of the visible and somewhat comprehensible aspect of outer space – when it rises, the odd patterns on its surface are there for all to see, whether they coalesce into the silhouette of a rabbit or an old woman.

The adolescent and the very young adult understand it differently: if a man did actually alight on the moon, and even brought back rocks that geologists could study, then there was something to be said for the exploration of space. Often, all that young minds need is a spark that can ignite their thinking and their innate curiosity. For the little boy from Kanpur, the tiny piece of moon rock proudly exhibited in Phool Bagh was that spark: he grew up to become one of the authors of this book.

Who knows what sparks of imagination will fly with Chandrayaan-1? A whole generation of kids may want to become astronauts – or antarikshayatri, if one may coin an Indian word. Outer space has caused major waves in the collective Indian psyche in recent times. It is not easy to forget how the country united to mourn the death of Kalpana Chawla, the Indian-born American astronaut who was killed in the Columbia disaster of 2003. And then came admiration and adulation for Sunita Williams, whose father's ancestral roots lie in Gujarat. Williams displayed astonishing skill in outer space – setting two daring world records for the longest stay in space (195 days) by a woman and the longest total duration ever of a space walk by an astronaut – during her flight to the International Space Station (ISS) in 2007. And who can forget India's favourite space hero, Rakesh Sharma, who flew into space in 1984 and declared India to be '*Sare jahan se achha*' as he looked out of a Russian space capsule.

This is an exciting time for space science in India. Six decades into independence, it is reaching for the skies in every sense of the word. The country was barely a decade into free governance when the sojourn began, and already we are viewing a significant gear shift – from application entirely for social development to the cutting edge of scientific exploration. This is a shift that could prove vital for young people who seek knowledge along with excitement, and also tangible opportunity.

With its first scientific mission to the moon, Chandrayaan-1, India has underscored its intent to progress from serving society to taking up the challenge of resolving the complexities of space science. Considering that it is often the other way round, there is global recognition that the Indian way has been a rather unique trajectory for any space programme.

This book is in some ways a celebration of this uniqueness and the far-sighted decisions that have led the country to this juncture. It is also about the marvels of the moon and the scientific curiosity it has generated in a developing nation populated by a billion people and grappling with nearly as many problems. In the past, lunar missions have succeeded in galvanizing entire nations and igniting collective psyches, and the Indian mission, in all likelihood, will do the same.

For there is absolutely no doubt that India is raising the ante on its space programme. The space community, with able support from the political leadership, is propelling the country towards ambitious missions to the moon, Mars and beyond. India's maiden mission to the moon in 2008 is separated by a mere four decades from the beginnings of the space programme. At that stage, Dr Vikram Ambalal Sarabhai, father of the Indian space programme, had said, 'In India, the immediate goals of our space research are modest. We do not expect to send a man to the moon or put elephants, white, pink or black, into orbit round the earth!'

Sarabhai's focus and seemingly self-effacing vision were critical to the creation of a strong foundation for the space programme – as will be evident in this book. Despite the borders he drew, Sarabhai was a dreamer. And his enthusiasm was infectious – whoever he worked with or whose lives he touched, the dreams seemed to live on in them.

In the past three decades, 26 large Indian rockets have been put to use for space launches, 3,000 small sounding rockets have been sent out, and 46 Indian-made satellites have gone out into space. Now, the country is seriously considering sending an Indian into space by 2015, and to the moon by 2025. The chairman of the Indian Space Research Organisation (ISRO), Dr G. Madhavan Nair, said in 2007: 'As far as humankind is concerned, yes, we have lots of problems to solve on planet earth. But, at the same time, the next frontier for humankind is going to be outer space. For various purposes, whether scientific observations, manufacturing, power generation or even from the security angle, outer space is going to be the most important location. So our presence in outer space will become a logical extension of any space research activity we undertake. And if we say that we are going to maintain a sort of leadership in space research, naturally the exploration of space, outer space, with human beings, will become a part of that. So with this backdrop we try to see what needs to be done in case you have to have a manned mission.'

◆

11 May 1999 was a big day for India, particularly for India's nuclear science community and for the government. A year earlier, India had joined the big league and blasted its way into the exclusive club of nations that possesses the technical know-how for nuclear bombs. The country's prime minister at the time, Atal Behari Vajpayee, had declared the date of the successful Pokharan-2 blasts as Technology Day.

Expectedly, there were several official functions and events going on in New Delhi, and there was a general sense of excitement. Most journalists spent the day looking for that one piece of information that would make it special. But at the end of three events, it seemed as though there was nothing out of the ordinary. The last thing on the agenda was a lecture in the evening, the Technology Day talk, to be delivered by the chairman of ISRO, Dr Krishnaswamy Kasturirangan.

Over 500 people were gathered in one wing of the sprawling Convention Hall in the somewhat grungy Ashok Hotel in New Delhi. The other wing was packed with exhibits of domestic technology developments. In his fluent hour-long lecture titled 'India's Space Odyssey', Kasturirangan spoke at length about India's capabilities in space, using a splendid PowerPoint display to reinforce his speech. Though long and detailed, the talk was largely devoid of anything novel, at least from the point of view of a hard-nosed science journalist. As the talk progressed, so did the angst. It seemed suddenly like a day wasted, spent as it had been, covering a mundane government event. But didn't somebody say patience is a virtue? Suddenly, and quite unexpectedly, in a blink-and-you-miss-it moment, late in the evening, everything changed. It was like being jolted awake, half-comprehending, by the alarm clock in the early hours of the morning. On the screen was a slide that, if read carefully, indicated ISRO's technical readiness to undertake a mission to the moon.

ISRO sending a mission to the moon? It was difficult to grasp the true import of this paradigm shift from application to exploration, from service to science. After his talk, in a brief chat, Kasturirangan provided some sketchy details of what he and his team at ISRO foresaw of an unmanned mission to the moon.

Chandrayaan-1 had been birthed.

The first call was made: ISRO had just announced its plans to visit the moon, and it would be difficult to say when one would get home. The second call was made to the forever awake and effervescent executive editor of the *Indian Express*. The response was immediate and an excited Raj Kamal Jha – a mechanical engineer from IIT by training – said, '*Aa jao, yahin karenge story. Lagayenge*,' which is best translated as 'Come right over, we'll do the story at the office. We'll place it'. By the time one reached the sprawl that is the newspaper's editorial office in the basement of the Express Building at India's Fleet Street, Bahadurshah Zafar Marg in New Delhi, it was past midnight. It felt strange, keying in the story with Atex software on an obsolete computer system, a battered machine which had for some reason a hole in place of its hard disk. Jha joked that it was a true 'Windows' system: it had a window!

The headline in all 17 editions of the *Indian Express* the next morning was 'Destination Moon – ISRO's Next Big Thing'.[1] Kasturirangan was quoted as saying that if all went well, the moon mission would be a reality by 2008.

Destination Moon is ISRO's next big thing

PALLAVA BAGLA
NEW DELHI, MAY 11

TODAY may have been the anniversary of Pokharan II but what caused more excitement in the scientific community was chairman K Kasturirangan's announcement that ISRO's Polar Satellite Launch Vehicle can "undertake a mission to the moon." And a core team of scientists is being put together to work out the details.

In his Technology Day lecture here this evening on "The Indian Space Odyssey," Kasturirangan said that India could easily launch a small satellite of about 275 kg in a "fly-by mission" to the moon or even place a 140-kg satellite in an orbit around the moon. The mission: to study the moon's core. A manned mission, however, is still far away.

Destination Moon, he said, could symbolise the next big challenge for ISRO which has satellite technology well under its belt.

Working out the mission's objectives and payload could take time and if all goes well, it could be a reality by 2008.

The launch vehicle "will not be the problem" he said. The cost will be estimated, he said, once the scientific details have been worked out and the government will be approached for funds, Kasturirangan said. According to ISRO's plan, the Indian way to go to the moon could be by injecting a satellite that has a lot of onboard fuel into a lunar transfer orbit and then using onboard rockets to nudge the satellite closer to the moon. Later, there could also be a moon landing.

Indian Express, *12 May 1999 (front page)*

Nine years after the story was penned and published, Kasturirangan must be a happy man. The timeline he presented has been adhered to and ISRO has delivered. Without major cost or time overruns, Indian scientists are undertaking the country's maiden voyage to the moon. And even now, Kasturirangan is generous in his acknowledgement of that story in the *Indian Express*, saying it had a serious countrywide impact. For people began to see that India could go to the moon.

Today, with the increasing number of missions to the earth's nearest neighbour, it seems as though the moon is, for the second time in history, all the rage. China and Japan launched missions in 2007, the US is planning one, and for India 2008 is *the* year. Three other nations are launching moon missions in the not-so-distant future.

One of the most exhaustive lunar exploration missions in the past three decades, Chandrayaan-1 is a major milestone in India's already mature space research programme. Soon after Sputnik-1 was launched in 1957 by the USSR, India set about addressing its own need for a home-grown programme. Today, together with China and Japan, India is part of a fast-developing Asian space sector. Even as Chandrayaan-1 awaits activation, India's future plans include a soft lander and rover on the moon by 2012, a manned mission in space as early as 2015, a manned landing on the moon possibly by 2025, and the use of the moon as a staging point for missions to Mars and beyond.

There is clear motivation and zeal for India to be part of the world's foray into to outer space. But where is all this zest and zeal coming from? There are over 300 million people living below the poverty line, a grossly inadequate education system, unemployment and malnutrition, and a large health burden: why would such a country undertake what seems to be a profligate scientific mission? Prime Minister Manmohan Singh addressed this question when he spoke out in defence of Chandrayaan-1: 'We have to walk on two legs, to deal with the fundamental problems of development and at the same time set our sights sufficiently high so we can operate on the frontiers of science and technology. In the increasingly globalized world we live in, a base of scientific and technical knowledge has emerged as a critical determinant of the wealth and status of nations and it is that which drives us to programmes of this type.'

Dr Manmohan Singh is not alone in feeling this way. Political leaders and academicians, physicists and citizens, felt the same 60 years ago, when India was undergoing the complex and challenging birthing phase of its now vibrant and diverse space research programme. In 1968, Sarabhai said: 'We are convinced that if we are to play a meaningful role nationally and in the community of nations, we must be second to none in the application of advanced technologies to the real problems of man and society, which we find in our country.' Today, with space research and technology touching our lives in many small and big ways, Sarabhai's testimony stands vindicated.

It goes without saying that India sees its mission as a unique response to a global challenge. Self-reliance is a serious driver after all, and it makes nations do several things, some routine and some unusual. Also, every country has a natural growth curve – once there is economic stability, you can be more daring with your national programmes and there can be a broader margin for risk taking. Kasturirangan has eloquently explained the force that drives India towards the moon and beyond: 'It is not a question of whether we can afford to go to the moon. It is whether we can afford to ignore it.'

That there are risks is clear as day. But just think how exciting it would be to obtain a 3D atlas of those regions of the moon that hold scientific promise. And a chemical map too, through an investigation of the distribution of various minerals and chemicals on the moon. In the words of Madhavan Nair, 'The objective of the lunar mission is to derive maximum scientific knowledge about the moon.'

Expected to cost approximately Rs 400 crore, Chandrayaan-1 is an unmanned orbiter or lunar satellite. The Polar Satellite Launch Vehicle (PSLV), India's locally-made 44.4-metre, 316-ton workhorse, in a new and improved avatar, is the launch vehicle for the moon probe that weighs about 590 kg in the moon's orbit and has space for scientific instruments of about 100 kg.

The moon probe is expected to last two years or more; most other missions have had much shorter lifespans. The Indian payloads will map the lunar topography and conduct X-ray and gamma-ray spectroscopic studies from a distance of 100 km from the moon's surface by placing a remote sensing satellite in a polar orbit.

Even as the mission was announced in 2003 and preparations were on in full swing, on a special request from President Dr A.P.J. Abdul Kalam, a 25-kg lunar impact probe or impactor was added to the mission. This tiny free-falling rocket, carried piggyback on Chandrayaan-1 can, on command, will detach from the satellite and hit the moon's surface, but it won't be a soft landing, as it is designed to strike the lunar surface at high speed. The probe is likely to throw up lunar dust, which would be instantly analyzed and the chemical composition beamed back to the orbiting satellite.

More than the value of the chemical data, the impactor will ensure that India's tricolour, which it carries, will be placed for eternity on lunar soil, making India one of the few countries, (the others being the US, Russia, and possibly Japan and the European Space Agency or ESA) to attempt placing their flag on the moon's surface. If the day comes when the spoils of the moon are shared among the nations of the world,

India cannot afford to be left out. And the Indian tricolour will be a reminder of its claim.

International collaboration and partnerships have always characterized scientific research in India, and the moon mission is no different. While planning the mission, the space research community in India had earmarked 10 kg for payloads from other countries. And so, payloads from NASA, ESA and Bulgaria are on board Chandrayaan-1. The significance of technology from the West travelling on missions from countries like India cannot be overstated, given the post-Pokharan sanctions. And then there is India's image of a not-so-glamorous scientific community, sometimes seen working out of laboratories housed in tin-roof sheds. It is indeed a great reversal of roles – India flying for free instruments fabricated in Western countries. As Dr Michael Griffin, administrator, NASA said: 'India and the US are not racing to go the moon. We can hope to go together.'

Chandrayaan-1 will place India squarely in the galaxy of scientifically advanced nations that seek to answer the unanswered, and explore unexplored frontiers. It will make India part of a small and select group of countries that have the capability to observe planets directly and with scientific rigour. The moon has always held pride of place in the Indian ethos for millions of people of diverse religions, communities and cultures across the length and breadth of the country. The twenty-first century will enrich this relationship by adding a whole new dimension of scientific discovery. After all, there are still many questions to answer, about the true origins of the moon and the possibility of commercially exploiting its natural resources in time to come.

To help achieve this goal of space exploration, the Planetary Sciences and Exploration Programme (Planex) was launched in October 2001 to give a fresh impetus to the creation and strengthening of research groups in planetary sciences in countrywide academic and research institutions. For, Planex, Physics Research Laboratory, Ahmedabad, is a nodal institution meant to build cadres of young space scientists who could directly be inducted into the varied dimensions of a lunar mission and other space research initiatives. Meanwhile, a programme called Research Sponsored (RESPOND), run by ISRO all over the country, is meant to provide financial support for research and development programmes in space science, technology and application. While focusing on encouraging top-of-the-line research, RESPOND also works towards building partnerships through which relevant research can be conducted and its results fed to support larger initiatives within ISRO.

People form the life and blood of any programme and India knew even in the days of Sarabhai that it needed more skilled people for this job. Without strategic focus on building human capacity, there could be trouble ahead. In 2008 a few radio stations repeatedly aired news of a contest for youngsters in which the prize was a trip to NASA. Listening to the announcement while driving to work, one wondered whether it was a real contest; if so, chances were it would hook at least one kid for life. What is it, after all, that attracts young people? Opportunity, being valued, and knowing that their skills can make a difference to the nation.

That said, it is critical to help increase the number of young people who take up space science as a discipline of study and also stay on to contribute through their research and active involvement with the Indian space programme. The Scientific Policy Resolution of 1958 had clearly committed itself to looking after those who offer their commitment to the development of science by offering good conditions of service to scientists and according them an honoured position in society, and it is important to keep that promise. India's entry into the elite lunar club of the world is a triumph; keeping the membership will be the real trial by fire.

1

Reaching for the Unknown: The Moon's Origins and the Search for Water

We make our world significant by the courage of our questions and by the depth of our answers.

– Carl Sagan, quoted in *Time*, 1980

The moon has an unquestionably special place in human hearts because of its links with diverse cultures and religions. As a repository of the geological processes that occurred during early planetary evolution, it is also an object of abiding scientific interest. When the first samples of moon rock were brought back to earth and analyzed, they were found to have formed 4.5 billion years ago. Here then, is a member of the space family that's been around for as many years, almost undisturbed, and is now like a time capsule, with a chronological etching of events that may have occurred in the early solar system.

Unlike planets such as Earth and Venus, where dramatic geological processes are still very active, the moon stopped evolving billions of years ago. This suspended state is critical to a broader and more accurate understanding of planetary evolution in space.[1] In short, if you have to understand the solar system, the moon is a good place to start.

The Moon We Know...

If the sun is the most prominent member of the solar system – the biggest and the brightest – the moon isn't far behind. It is the second brightest, and the fifth largest, satellite. Rocky and dry, and filled with dust, the moon has a diameter that approximates the distance between Srinagar and Kochi, which is about 3400 km. The systematic study of the moon, initiated in 1799, was termed selenography. In the early part of the seventeenth century, Galileo used a telescope to observe the surface of the moon, and noted the lowlands and the highlands. Lunar photography became known only by the end of the following century. Observers and astronomers in ancient times used to describe the dark patches on the surface of the moon as seas, called *maria* in Latin.[2] The other clearly demarcated area on the moon's surface is the mountain-like region, the highlands.

The moon has no atmosphere, and therefore no surface water that is stable in the way water on earth is. The side of the moon on which the sun's rays fall has temperatures that can touch 110 degrees Celsius; the side away from the sun can freeze to –180 degrees Celsius. Areas near the moon's poles are always dark and cold, with temperatures plummeting to –230 degrees Celsius, since the sun's rays never reach there. These extremes of temperature certainly don't make it a very livable place, but studies from the Lunar Prospector, an American spacecraft that went up in 1998, suggest that there might be water ice on the moon.

Being so close to the earth, the moon wields a great influence over life on earth and its evolution, which makes it more of a scientific curiosity than other planetary bodies.[3] Scientists have always been keen to learn as much as possible of its physical and chemical aspects. Interestingly, processes that are supposed to have modified some of the planets of the solar system have not been identified on the moon. For instance, plate tectonics have churned the earth's crust continuously; wind and water have altered the Martian surface a number of times.[4] Since these changes have not occurred on the moon, its core is understood to be a faithful record of the evolutionary stages of the solar system.

In 1970, scientists at the Tata Institute of Fundamental Research (TIFR), Mumbai began direct analysis of moon rock samples available after the Apollo 11 mission. Dr Devendra Lal and Dr Narendra Bhandari were the first Indians to hold moon rocks in their hands and analyze them. Lal recounts an interesting historical fact – India was the first country with which the US shared moon rock samples. Bhandari himself had worked in the US at a time when the Apollo missions were being conceived, but returned to continue his exploratory analysis of lunar materials in India. He recollects how it took some time, after the Apollo 11 samples came in 1969, to put together a scientific proposal and receive approval for the work from NASA. He had, with Lal, a joint proposal on cosmic ray track studies, and a separate one on radioactivity measurements – these were the two aspects of planned research at TIFR. An entire team was involved in taking forward this research with Lal and Bhandari, including V.S. Venkatavaradan, Suman Bhat, G. Rajagopalan, and A.S. Tamhane. They worked on samples from both Russian (Luna) and American missions, trying to measure and analyze the effects of cosmic radiation and radioactivity in the rocks.

Dr Narendra Bhandari, Lucknow, 2000

Ever since lunar exploration began, orbiters, landers and rovers, even missions that would return samples to earth, have helped piece together our knowledge of the moon. We now have a general idea of the moon's vital statistics (see Table 1.1) and a calendar of major geological events that led to the early evolutionary stages of the moon, the formation of its crust and interiors – but its origin is still a mystery.

How did the moon come into existence? The most popular theory is the Giant Impact Hypothesis, which holds that a celestial body about the size of Mars collided with the earth soon after it had formed, and the impact was such that a large part of the earth was blasted into space, with debris floating around. Some of this debris (or ejecta) started to orbit the earth and coalesced into the moon. This is a popular hypothesis because it is prima facie very plausible. For instance, the material that got blasted off contained very little iron, which would explain the lack of iron on the moon. Also, moon and earth rocks are very similar and the two bodies have tilted orbits.

There are other theories too. Was the moon formed along with the earth – the theory of co-accretion or simultaneous formation – out of solar nebula? While this theory passes muster if based on the fact that there are similarities between the rocks of the two, it isn't entirely convincing because it cannot explain why the soil on the moon is low in iron. Another theory, the Capture Theory, postulates that the moon was originally a planetary body that formed elsewhere in the solar system, and was captured into orbit as it came close to the earth. Then there is the Fission Theory: while the earth was spinning, a part of it separated, cooled and became the moon. The laws of momentum and viscosity, however, go against this theory.

While the search for consensus on the origin of the moon continues, missions and studies over several decades have resulted in a map of the region. The highland crust clearly has three zones – the South Pole-Aitken basin (this is the largest basin on the moon or on any planet); the feldspathic highlands and Procellarum-KREEP.[5] Among the most fascinating areas is the South Pole-Aitken basin, 12 km deep and 2,500 km in diameter. Since it seems to have been created by some massive impact, experts feel it may well be an indicator of the geological nature of the moon's mantle.[6] There is a small impact crater in this basin that is named after Dr Homi Jehangir Bhabha, father of the Indian atomic energy programme, close to the J.C. Bose crater – these are on the moon's far side, on its southern part. There are also craters named after Sarabhai and astrophysicist Meghnad Saha, which are on the near side of the moon.

...And the Moon We Don't Know So Well

The iconic British band Pink Floyd sang, *'I'll see you on the dark side of the moon.'* But beyond artistic liberty, Bhandari and other experts explain how, given the identical period of the moon's rotation and its orbital movement, we see only one side of the moon from the earth. The other side, therefore, is often referred to as the dark side of the moon. In fact, experts say that we know the surface of Mars better than we know that of the moon, despite the large number of missions, the samples brought home and centuries of study through telescopes. Even as knowledge about the moon reaches a scale and dimension that makes it the best understood planetary body for humankind after the earth, there is a concurrent need to address its many, many unknowns. As Prof Bhandari and his colleagues have said, 'We still do not know enough about the moon's formation, its internal structure, core and magma formation, chemical and mineral stratigraphy.'[7] But there is no doubt that the moon is 'priceless to planetary scientists'[8] and may hold more answers than humankind may be prepared for.

So, is there water on the moon, maybe in the form of ice? Studies from the Lunar Prospector in 1998–99 showed that a number of neutrons of different energy levels coming out of the polar regions were peculiar compared to the rest of the moon, which meant the possible existence of hydrogen, which in turn meant there must be water. But these experiments were not confirmatory in nature, and the search is still on. The point is simple: if there is water or ice on the moon, it would make for cost-effective habitation[9] as the world plans other exploratory trips using the moon as a base. Then again, it isn't that simple. Studies such as those conducted by lunar missions such as Prospector and Clementine (the latter used radar reflectivity experiments to raise the possibility that there were large sheets of ice in the shadowy regions), have not been conclusive. Originally, the possibility of water on the moon was raised because there are several planetary objects like comets and meteorites that crash into the moon off and on, and many of these have water. It is possible that they have left behind water molecules on the surface of the moon. Of course, the sunlight and heat would cause most of it to evaporate, but since the moon's axis is only slightly tilted, there are several craters on the poles that are in permanent shadow and may become natural traps for water. For now, humans are still looking for concrete evidence of water on the moon.

Is there another body out there that is like the moon? Or is it unique? This is a question we can't answer yet, but what we have learnt so far is that the moon is a mirror to the history of the solar system, which goes back 4.5 billion years. As a report of the National Research Council put it, 'Nowhere else can we see back with such clarity to the time when Earth and the other terrestrial planets were formed.'[10]

It was in 500 AD that Aryabhata determined the size of the moon and its distance from the earth. The quest that he and other Indian astronomers started continues 1,500 years later, as we prepare for lunar missions that we hope will help fill at least some of the gaps in our knowledge of the lunar world.

Table 1.1: Vital Statistics of the Moon

(Source: Wikipedia and Dr Bhandari)

Circumference	10,921 km (equatorial)
Surface area	3.793×10^7 km^2
Volume	2.1958×10^{10} km^3
Mass	7.3477×10^{22} kg
Temperature	-230 degrees C at poles;
Range	-180 degrees to +110 degrees C during night and day

2

The Moon in the Indian Ethos

Chanda re, chanda re, kabhi to zameen par aa, baithenge batein karenge;
Tujhko aate idhar, laaj aaye agar, odh ke aaja tu baadal ghane
(Moon, O moon, do come down to earth sometime, we'll sit and talk;
If you're shy coming down here, just wrap some thick dark clouds around you)

– Javed Akhtar

'Chanda' has lent itself to some of Hindi cinema's most memorable numbers. From the early days of Guru Dutt serenading Waheeda Rehman with '*Chaudhvin ka chand ho*' to '*Chanda re*' in the film *Sapney*, the moon has been a recurring theme in Indian cinema. In an industry where the expression of love has to rely on imagination and innovation rather than physical overtures, the moon offers a safe alternative. Hindi movies of old had a stock scene: a full moon would drift out from behind a dramatic-looking tree, watching benevolently over the romantic pair as they went about singing and frolicking.

The moon is also popular with children – some see a rabbit on the moon, and others see a cuddly granny. Hindi movies don't forget children either, creating lullabies every now and then, the most famous being the timeless '*Chanda mama door ke*'. The world of music has always had a special bond with the moon, across genres and cultures. Beethoven's Moonlight Sonata is one of his most memorable compositions. In fact, when he composed it in 1801, Beethoven called it '*Quasi una Fantasia*', Italian for 'Almost a Fantasy'. But after his death, the poet Ludwig Rellstab renamed the first movement of the piece Moonlight Sonata. Listening to the composition, he said, was like seeing moonbeams shining on Lake Lucerne.

Then there is the Hindu inerpretation of the birth of the moon. The dramatic Kshirasagar Manthan, or the churning of the primordial ocean of milk, is among Hindu mythology's most theatrical events. It is also central to the religion and its mythical beliefs. When the devas and asuras desired amrit, or the nectar of immortality, they decided to churn up the ocean of milk. The story goes that in the process of churning, a deadly poison or halakalavisha was thrown up and there was fear that it would destroy everything. The devas appealed to Shiva, who acted quickly to swallow the poison. When the churning began again, the moon came up from the ocean – ethereal, luminous and beautiful. To express their gratitude to Shiva, the celestial witnesses to the event took a collective decision that the moon should become a jewel on his head, and that is how the crescent moon came to adorn the Destroyer's matted crown of hair. If Shiva was honoured to possess this new and strange adornment, it was an even greater honour for the moon to gain pride of place on the person of Nataraja, or the god of the cosmic dance. It is believed that the Tandava Nritya arouses cosmic energies sufficient to shape the world and that the moon was granted cosmic divinity after being placed on Shiva's head.

Many religions interpret the cosmos as a result of divine action. Perhaps it is just human instinct – to try and find reason and logic for everything we see and experience. This hunt for patterns often extends to relating planetary bodies and their complex and diverse movements to earthly activities, thus making sense of that which is beyond human comprehension. It is also a way of reaching out to objects animate and inanimate, entwining them into our daily routine, and using their comforting mythic nature to explain the inexplicable. This is how planetary bodies – distant though they may seem – have become part and parcel of daily life on earth. Incredibly, this is true across communities, nations, cultures and faiths.

Of Gods, Goddesses and Mythical Tales

The moon is worshipped in many parts of the world, in many cultures. Not only is it revered and paid obeisance to, but the gods and goddesses representing the moon are worshipped as well. In Hinduism, Soma is the moon god, galloping through the clouds on a gorgeous chariot drawn by white horses. Soma is said to be the one who provides the nectar of life to the gods, and the moon is where this elixir is stored. Believed to be the son of Varuna, or the god of the oceans – where Soma was birthed in the churning – the moon god carries a mace, a pitcher of amrit and a lotus flower. With all the gods drinking nectar, Soma periodically becomes weak, which explains the waxing and waning of the moon. Lakshmi, the Hindu goddess of divinity, prosperity and purity, also emerged from the ocean and chose Vishnu, the Preserver, as her consort. Lakshmi and Soma are considered to be sister and brother, since they emerged from the same ocean.

In the ancient Greek tradition, Selene was the beautiful goddess of the moon, pale-faced and delicate. Interestingly, her vehicle too is a chariot of silver, pulled by horses, or sometimes a pair of oxen. Selene, with a half-moon on her head and carrying a torch, is a popular depiction in post-Renaissance art in Europe. In Latin, the same goddess is called Luna. The Temple of Luna – dedicated to the moon goddess – was destroyed during the Great Fire of Rome.

For the Inuit of Greenland, the moon god is male and is called Anningan. He is the brother of Malina, the sun goddess. The story goes that the two were happy together until Anningan became jealous of his sister's beauty and attacked her. A seal-oil lamp tipped over and Malina, her hands black with grease, retaliated by sullying Anningan's face. As the fight worsened, Malina escaped far into the sky, and became the sun. Anningan chased her and became the moon, and that is how the eternal race between the sun and the moon began. When a solar or lunar eclipse occurs, the Inuits explain it as Anningan catching up with Malina and attacking her again; when the moon wanes, they say it is because Anningan's anger with Malina is on the rise, which makes him go off food and become thinner.

In China, the moon festival is also the mid-autumn festival. The harvest moon – which appears on the fifteenth day of the eighth lunar month – is considered to be the biggest and brightest full moon of the year. On this night, people carry paper lanterns as they step out to sight the moon, and there is celebration and merriment. There are various interpretations of the legend behind this festival. One of them is the story of Chang'e, a beautiful girl who married an archer and hunter called Hou Yi. Chang'e once mistakenly swallowed a pill containing the elixir of immortality that was meant for Hou Yi. To escape his wrath, she flew off to the moon and has been there since, and it is her beauty that is reflected in the lovely harvest moon.

Among the aboriginal communities of Australia, people believe that a sun-woman awakens each day in the east. She carries with her a brightly burning torch of bark and she signals to her people that it is time to leave home and hunt. As evening approaches, a moon-man makes an appearance – with a smaller torch of light – and moves across the sky, illuminating the night. This understanding of the sun and the moon is based on a legend in which Purukupali and Japara – the earliest men on earth – rubbed sticks together to discover fire. Aware that this was a precious discovery and they could now actually cook food, Purukupali handed a burning bark to his sister, Wuriupranala, and a smaller one to Japara. His only request to them was to keep the fire burning. The sun and the moon are proof of the fact that they did so.

The Moon in Our Lives

It is hard to imagine the existence of human society without a calendar: astronomical calendars are a cornerstone of human life. There are three types of calendars – the solar, the lunar and the luni-solar. The solar calendar, also called the Gregorian calendar, which the Western world follows, goes by the tropical year (the time it takes for the earth to go around the sun once), but it needs the leap year with its extra day every four years to catch up.

The lunar calendar is based on the moon's movements; Islamic societies use a purely lunar calendar. A lunar calendar can have varying start-days each month, depending on the presence of the new moon in a particular time zone. Each lunar month lasts 29-30 days and each day is called a 'tithi' at sunrise.

The Indian calendar is structured on both the sun and the moon – the year is solar, but divided into 12 lunar months. In the Indian calendar, seasons follow the sun; months follow the moon; and days, both the sun and the moon. For the Hindus, the 12 months are Chaitra (starting mid-March), Vaishakh (mid-April), Jyaishtha (mid-May), Ashadha (mid-June), Shravana (mid-July), Bhadrapad (mid-August), Ashwin (mid-September), Kartik (mid-October), Margashirsha (mid-November), Paush (mid-December), Magh (mid-January) and Phalgun (mid-February). These months are divided into two halves of 15 days each – the shuklapaksha or the light, waxing phase, and the krishnapaksha, or dark, waning half. Even within Hinduism, there are two ways of interpreting and using the lunar calendar. In southern India, the amanta or mukhya mana system is practised, with the month beginning with a new moon. In the north, the purnimanta or gauna mana system is prevalent, according to which the month begins with a full moon.

For many Indian children, there is great excitement in running out to spot the delicate crescent moon – duj ka chand – for their grandmothers. Pallava, one of the authors of this book, remembers that his grandmother's religious rituals were synchronized with the moon in its different shapes –

spotting the tiny sliver even as the setting sun's rays were still trailing the sky meant her month had begun. There were no prayers offered, this was just the matter-of-fact marking of a new month. Ignorant of the date of her own birth on the Gregorian calendar, her life was ruled by the moon.

Practising Hindus place their birth dates according to the 'nakshatras' or the 27 lunar mansions – divisions in the sky identified by the presence of prominent stars in each – through which the moon passes periodically. The Panchangam, the astrological almanac of the Hindus, is also based on Indian interpretations of cosmology, including the moon's movements. It is this almanac that is referred to for all traditional events and dates. Within Vedic astrology, the moon is considered to be part of the Navagraha, along with the sun, Mars, Mercury, Jupiter, Venus, Saturn, and Rahu and Ketu – the latter two being the northern and southern nodes of the moon. In astrology, they are interpreted as invisible planets which, at certain times of the year, swallow the sun or the moon, causing a solar or lunar eclipse.

Celebrating with the Moon

Devout Hindu women in several states of northern India fast once a year for the well-being of their husbands. Karva Chauth is a special day for a lot of people – and while the day is spent fasting, there's dressing up and celebration towards the evening. The high point of the day is when the fasting woman gazes through a sieve, first at the moon and then at her husband, after which the fast can be broken. This fast is observed in the Hindu month called Kartik (October–November), nine days before Diwali, on the fourth day of the waning moon. This is one Hindu observance that makes the moon the centre of attraction – it must pop out and be seen so the hungry women can finally eat. And of course, the husbands must oblige too.

What would our cultural practices be without the moon? Most Indian festivals and auspicious days don't recur on fixed days of the Gregorian calendar – they follow the lunar calendar. Each year there is a minor variation in the date when Diwali or Rakshabandhan or Eid is celebrated. This is because the luni-solar or lunar calendar is followed. So, the full moon day in the Hindu month of Shravan, which falls between August and September each year, is celebrated as Rakshabandhan. Similarly, on the new moon in the month of Kartik (around November), Diwali is celebrated. Eid too is celebrated on sighting the moon, in its delicate crescent shape towards the end of Ramzan. Meanwhile, twice every month, on the eleventh day of the lunar cycle, Hindus observe Ekadashi. According to Hinduism, the moon is the presiding deity of the human mind, and the observance of Ekadashi is in some ways celebration of the relationship between the two.

The moon's movements are also significant in determining the auspicious days for bathing in the Ganga at the world-famous Kumbh mela. The origin of the Kumbh is again linked to the Kshirasagar Manthan – the story is that the devas and asuras fought for 12 days and 12 nights (considered equivalent to 12 years for human beings) over the immortalizing nectar. As they fought, drops of nectar fell from the pitcher at four places – Prayag or Allahabad in Uttar Pradesh, Nashik in Maharashtra, Ujjain in Madhya Pradesh and Haridwar in Uttarakhand – which is where the Kumbh is celebrated.

While the Kumbh is celebrated once in 12 years, the Ardha Kumbh is celebrated every six years. The dates of the mela are based on the relative position of the sun and Jupiter. The Maha Kumbh mela (which takes place once in 144 years) of 2001 is considered to have attracted the single largest congregation of human beings ever on this planet. The mela is spectacular for its scale and its unique display of spiritualism, faith and monastic practice. People also journey to the banks of the Ganga and other holy rivers in India to bathe on certain auspicious days (such as the Paush Purnima or full moon day, and Mauni Amavasya, moonless night) and wash away their sins.

A full moon – when the moon is on the side of the earth exactly opposite the sun and the three are completely aligned – is what determines a host of rituals and festivals. In Mathura and Vrindavan – bustling, crowded towns in northern India better known as the ancient land of Krishna – Sharad Purnima, a full moon day in October when the moon is understood to be closest to the earth, is a special day. The colourful Raas Lila – the romantic dance performed by Krishna, Radha and the gopis – apparently took place on Sharad Purnima and the day is associated with love and romance. On the thirteenth day of the rising moon is when Mahavir, the holy saint who founded Jainism, one of India's major religions, was born, in about 600 BC. Mahavir Jayanti is celebrated in the lunar month of Chaitra (March–April). Again, it is a full moon day in the month of Kartik – Kartik Purnima – when the Sikhs celebrate the birthday of their founder saint, Guru Nanak. For the Buddhists too, the full moon holds a great deal of significance. On that day in the month of May, Buddhists worldwide celebrate the birth, enlightenment and Nirvana of Buddha.

For the longest time, it was understood by psychologists that the full moon influenced human behaviour, affecting mental balance and causing crimes to be perpetrated. That is the origin of the word 'lunatic'. Modern psychiatry does not give any credence to these studies – but beliefs know no data, as they say.

A lunar eclipse also holds meaning for believers. In Islam, for instance, a special prayer offered during the solar/lunar eclipse asks Allah to protect his followers and their souls made bright by the sun and the moon, from the spiritual darkness, which brings in its wake misconception, sin and evil. For practising Hindus too, there are several beliefs that must be adhered to during a solar or lunar eclipse: parts of it are considered auspicious and others are not. Food cooked during an eclipse is not to be consumed; it is common to observe a fast on this day. In Punjab, a pregnant woman has to stay indoors, not move, eat or change clothes. At certain holy places, particularly Kurukshetra in Haryana, considered the original battleground of the Mahabharata, bathing in one of the beautiful holy tanks of the city is supposed to bestow a thousand times more good on the devotee than performing an Ashwamedha Yagna (a ceremony meant to engender blessings). Interestingly, there is also good to be had from an eclipse, so any donation given during this period is considered highly auspicious.

Since the earth is tilted on its axis, the moon's gravity is felt varyingly during the course of a 24-hour cycle – the centre and far side of the earth feel the pull. As a result, the oceans stretch in response to the moon's gravitational pull, but in a sort of an ellipse, forming a bulge. High and low tide in the oceans – because of this gravitational pull on certain parts of our wobbly earth – have a huge impact on human life, especially for communities that live by the sea and use it as their source of livelihood.

In the words of Francis Thompson in *Sister Songs*: 'The innocent moon, which nothing does but shine, moves all the laboring surges of the world.' It is amazing how, just like the ocean tides that have such an impact on human life, the moon seems to exert its influence on all spheres of human life.

As Vivekananda said: 'This is not the moon at all, but the habitation of the Gods.'

3

The World's Many Missions

From the Present into the Past

I do not expect that the debate on the merit or otherwise of putting man into space would ever be settled. If we are to rely on historical experience, man will surely push ahead with adventures of this type backed by motives which will inevitably be mixed.

– Vikram Sarabhai

In the mid twentieth century, there was a lot of excitement about going to the moon. This, despite the fact that understanding and unravelling its many mysteries called for considerable technological power and investment, political backing and sheer innovation.

The exploratory honeymoon lasted 17 years, ending in 1976. Though several burning scientific questions remained unanswered, the general message doing the rounds was 'been there, done that'. Even so, those 17 years were challenging, and therefore immensely satisfying for the nations involved, considering they were creating history in the exploration of the universe beyond the earth, while simultaneously developing new technologies and capabilities. Many of these technologies (such as Velcro) also had vital spinoffs through their application for human society.

For at least two nations, this was a very busy phase. The Cold War played a critical role – some would say even a pivotal role – in prompting the USSR and the US to set out on one space mission after another, including some to the moon. These successive missions, extensively documented and peppered with varied accounts (not all accurate) shaped much of the early history of the exploration of the moon. Between them, the US and the USSR crammed 55 missions into the 17 years, half of them between 1965 and 1969, obsessed as they were with scoring off each other in the game of touching the moon and getting back. Though the profligacy of the race can't be denied, it does not diminish the scientific value of the missions or their contribution to our knowledge of the moon and the honing of technology for space exploration.

Then, as suddenly as it had begun, the race cooled off. After the giddiness of the initial triumphs, both the Russians and the Americans pulled back. As for the rest of the world, there was either no capacity or no marking of priority. So for 14 years, there was no traffic on the highway to the moon.

However, from 1990 onwards, lunar exploration was re-ignited and there have been several missions, nearly one in four years, led by Japan, the US, the European Space Agency and China. But this time around, it isn't a race. Rather, there is a desire to explore potential resources, including those for energy, even as the earth starts to sag under the load of humanity. In other words, most of these more recent missions have been undertaken to sharpen technique and for nations to develop self-reliance. The answers they say they seek, such as unravelling the origins of the moon or the physical and chemical character of its surface, are often secondary goals, which must not be read as being any lesser.

Missions of the New Millennium

The second race to the moon started with Japan's launch of Hiten in 1990. Hiten was a flyby – it flew past the moon to study it and conducted specific technological experiments; it circumnavigated the earth and the moon and finally, as per plan, directly impacted the southeast part of the lunar far side. Hiten's successor, Kaguya, is currently beaming back pictures of the moon. The satellite is named after a baby in a Japanese fairytale who is found in a bamboo clump by an old couple and turns out to be a princess from the moon. Formerly called Selene, Kaguya suffered many delays before being finally sent into orbit in September 2007 from the Tanegashima Space Centre.

Japan initiated its efforts in the 1950s, through test flights by a small rocket aptly christened 'Pencil'.[1] The Japanese Aerospace Exploration Agency (JAXA) is a recently consolidated agency that now is the sole institution in the country dedicated to all space-related activities. According to JAXA, Kaguya had obtained, by November 2007, continuous images of the far side of the moon: the 'world's first continuous reflectance spectra of the far side of the moon in the visible and near infrared region'.[2] On board the satellite was a piece of equipment called a Spectral Profiler that had managed to 'successfully produce a series of spectra along a strip longer than 1,000 km on the far side of the moon'.

Kaguya has one main orbiting satellite and two smaller satellites (Rstar and VRAD) in a 100-km polar orbit. Kaguya's normal operations kicked off in mid-December 2007 and formal scientific investigations of the moon began. Kaguya plans to verify certain scientific details, such as the status of the lunar magma ocean believed to have been formed by the impact that led to the formation of the moon itself, according to the Giant Impact Hypothesis. In fact, the idea that the surface of the moon was a thick molten mass that cooled off to form a crust several hundred kilometres in thickness fits with the geological nature of the '400 kilograms of moon rock samples that Apollo astronauts hauled back'.[3] At a cost of $458 million, Kaguya is expected to use 15 remote sensing instruments that will be directed towards the moon from a distance of 100 km. Observations and studies will be aimed at figuring out the tectonic history of the moon by analyzing elements and minerals and the way they are distributed on the lunar surface.

Around the same time, in November 2007, China's first lunar orbiter, Chang'e-1, started transmitting photos of the surface of the moon. China's space programme was initiated in 1956 and saw its first major milestone in 1970 when the country launched its first man-made satellite. The Chang'e mission cost around $264 million and evolved after the manned mission programme was put in place. Like Kaguya, Chang'e is also named after a character in a folktale, an angel who drank a magic potion and flew away to the moon. Launched in October 2007 from the Xichang satellite launch centre in Sichuan province, Chang'e is studying the nature of the space environment between the earth and the moon, besides detailing characteristics of the lunar surface and its temperature.[4]

In the late nineties, Dr Luan Enije, administrator of the China National Space Administration (CNSA), and his team were trying to determine the direction the Chinese space programme should take in the coming years.[5] Although researchers were keen to build and implement a lunar exploration programme from the early nineties on, the government had prioritized a manned mission.[6] Enije and his colleagues proposed what was called the '211' development plan, meant to include two spectrums, both launch vehicle and satellite platforms, one satellite application programme and one programme for deep space exploration.[7] Chang'e indicated that China meant business when it came to the deep space exploration part of its 211 plan. The Chang'e satellite weighs two tons and is currently in a 200-km circular polar orbit. Its year-long mission is to carry out several studies elucidating the nature of the moon's surface, using three-dimensional imaging methods.

Unlike Chandrayaan-1, Chang'e has been flown without the participation of any other nations. China is a country in a hurry, as Dr Wu Ji, Executive Director, Centre for Space Science and Applied Research in Beijing, remarked in the course of an interview. The Chang'e project was started in 2004, some three decades after China's first steps in this direction, and it involved developing a fleet of launch vehicles and creating several satellites for various applications. Now, the Chinese space programme is set on placing a robotic lander on the moon in roughly five years, and following this up with a sample-return mission.

By September 2003, ESA launched the first of its Small Missions for Advanced Research in Technology (SMART-1) to validate solar-electric propulsion systems as technology for lunar missions. SMART-1 used a unique ion propulsion system with xenon gas, a key component for gentle propulsion into lunar orbit, and took almost a year to get to the moon. The 287-kg SMART-1 carried an entire battery of miniaturized instruments.[8]

By mid-November 2004, SMART-1 was in lunar orbit and mapped some of the elements on the lunar surface. The satellite carried out studies to try and validate one of the theories about the evolution of the moon – that a violent collision between the earth and a smaller planet caused its formation. It obtained maps of hitherto unstudied regions of the moon – such as the lunar north pole – before its planned crash-landing on the surface of the moon in September 2006. According to Bernard Foing, SMART-1 project scientist, 'Solar illumination makes these areas ideal for robotic outposts or lunar bases making use of solar power.'

SMART-1 was able to capture images of the Peary impact crater (very close to the north pole) when the crater floor was receiving sunlight. In fact, the fascination for the polar regions is exactly this – their illumination by sunlight. It makes them exciting regions for further exploration because they not only have what the experts call 'quasi-eternal' light, but are also close to darker regions that may have ice. All of this translates into useful knowledge that may later be used in the world's attempts at outer space exploration. In fact, these sites may well become a base for future missions.

Of course, the search for water ice on the lunar poles continues as SMART-1 could not come up with conclusive evidence. Experts such as Bhandari explain how SMART-1 was primarily a technological mission and that it was limited in its value addition to scientific information regarding the moon's chemical composition or the existence of water.

Back on the Radar

After the early seventies, when the US–USSR moon race wound down, it was Japan that broke the lull with Hiten in 1990. Clearly, the cooling-off period from 1976 to 1990 had proved useful – there was newer and more sophisticated technology, which gave the endeavour a fresh fillip.[9] Also, the Cold War was over, and the collaborative spirit among nations was growing. For instance, Clementine, a collaborative venture of the Europeans and Americans, used high-resolution cameras and laser image detection techniques to collect the first global data sets related to gravity and the moon's topography. This satellite also managed to observe the Peary crater and picked up areas with sunlight. The Lunar Prospector was launched in 1998 and carried out extensive chemical data collation, besides raising the possibility that there could be water on the moon, frozen and spread out on the shadowed polar regions.

The nineties saw little activity, as compared to the rush a few decades earlier. The writing on the wall, though, was clear – technology had improved and more nations were prepared to go lunar than ever before. Some experts called this the mini-renaissance of lunar studies[10]; it seemed as though the busy days of the lunar sixties were making a comeback.

Two Nations, One Race

It is possible to push the frontiers of space exploration today only because of the foundations laid in the early days of the space race. Sputnik-1 had opened the door to space, being the first to go beyond the earth's atmosphere and into space in the real sense of the word. That was in 1957. It spurred the US into action, and kickstarted a race between the two superpowers. Three Sputnik missions, and success – the Russians were gazing at the moon. The Luna family of satellites – a 24-part series – made an enormous difference to human understanding of the moon, and technology that would impact this understanding. The first lunar flyby, Luna-1, also became the first solar orbiter, but it was Luna-2, again a Russian spacecraft, that actually impacted the moon's surface. All of this happened in 1959. That the moon had no magnetic field or radiation belts around it was confirmed on this mission.

Vostok-1, a Russian rocket, took off into space with Major Yuri A. Gagarin in April 1961. Orbiting the earth for 108 minutes, the Russians managed to show the world that human beings could make it to space and back – and in the process they immortalized Gagarin's name. Managing a soft lunar landing was another milestone for the Russians – Luna-9 did this in 1966. Luna-16, a robotic probe, not only landed on the moon but also returned a sample to earth for analysis.

Russia's achievements motivated the US to go further. In the words of H. 'Jack' Schmitt, geologist, astronaut, senator and author:

> The initial catalyst for Americans venturing into deep space was the Soviet Union's October 1957 launch of Sputnik-1, the first artificial satellite of the Earth. It burst upon the American consciousness as one of the defining moments in the history of the United

> States. Although a temporary propaganda coup for the Soviets, the law of unintended consequences took over as a technological giant became focused on the obvious long-term importance of space. Thousands of young Americans began to think of space in the context of their personal futures and the future of the world and began to plan their education accordingly.[11]

There were 31 US missions between 1959 and 1976 as against 29 missions from the USSR. The missions mounted by the US, also in the late fifties and early sixties – Ranger, Surveyor and Lunar Orbiter – brought back photographic maps of the moon. In those early days, the level of excitement made it seem like science fiction. The US spent a lot of money and effort on the missions of the sixties and early seventies, inevitably raising questions about the need for such extravagance. To this day, doubts are cast on the utility of the Apollo programme: 'It will be helpful to look back at our first lunar program, Apollo, and ask what we got from it, besides some 850 pounds of rock and soil – fascinating to geologists, but perhaps not to all taxpayers.'[12]

With typical American extravagance, there were seven missions in 1967 alone, setting out to achieve varied goals, imaging among them. While the Apollo missions are the most talked about, a predecessor to these missions was Gemini, a technological warm-up, in NASA's own terms.[13] Gemini – with ten manned missions into space between 1965 and 1966 – managed to empower the US space programme with experience in space flight technique, and experimentation and research methods.

The Apollo programme – between 1961 and 1975 – was aimed at landing men on the moon. What is now the Johnson Space Centre came up as the Manned Spacecraft Centre, when the Apollo programme was first conceived. All things considered, the Americans progressed steadily with their lunar programme till they achieved their biggest win, when Neil Armstrong along with Buzz Aldrin set foot on the surface of the moon on 20 July 1969. The world sat up and took notice. Armstrong spent about two and a half hours on the surface of the moon, after having made his historic statement – 'That's one small step for [a] man, one giant leap for mankind.'

Timed as they were to the minute, Armstrong and Aldrin had to go through all the pre-determined procedures – including planting the US flag on the moon – with clockwork precision. After Armstrong, more than a dozen astronauts have journeyed to the moon and brought back materials for study. There were five Apollo missions after Apollo 11, and finally in 1972 came the last human landing of that phase, by Harrison Schmitt. The world-famous image of the earth called the Blue Marble was clicked on this mission, Apollo 17. In fact, among the most significant contributions of the Apollo programme was the images of the earth it provided – never-seen-before pictures of a fragile earth that needed care and attention. It is sometimes said that the early impetus for the global environmental movement to save the earth was provided by some of these telling images.

Today, Schmitt is best known for giving a practical, business-model-like spin to the human interpretation of lunar exploration. Early Apollo missions studied soil mechanics, magnetic fields and lunar ranging, and conducted multispectral terrain photography, among other things. The Apollo missions undoubtedly cemented the earth-moon relationship by emphasizing the centrality of humans in that magical bond.

Despite the success of all these missions over several decades, the generous allocation of resources by governments and the spectacular leaps in technological sophistication, there are still several unanswered questions. Many of the missions described above went searching for water or ice on the moon, but that still remains a primary research question. So is the need to identify potential landing sites on the moon. And then, of course, the more futuristic need to map its resources. Lunar missions across the world have gone step-by-step, slowly covering ground with logic and reason – from just flying past to sending robotic instruments down, to having humans land, to sending more and more sophisticated probes for increasingly complex experimentation and research. And the trigger for such action? A need to master technology, a nation's need to feel free to run along to the moon whenever the need arises!

Besides all of this, there's of course the element of a quest, as Madhavan Nair said in July 2007: 'Since the initiation of lunar exploration programmes as early as 1959, more than 100 exploratory missions have been carried out and a wealth of lunar data has been accumulated. However, the critical and fundamental questions about the moon's origin, formation and inner structure, chemical/mineralogical composition still exist.' In a nutshell, still miles to go.

Table 2.1: Chronology of Lunar Missions
(Source: www.isro.gov.in)

No.	Launch date	Mission	Country	Accomplishment
1.	2 Jan 1959	Luna 1	USSR	First lunar flyby
2.	3 Mar 1959	Pioneer 4	USA	Lunar flyby by 60,000 km
3.	12 Sept 1959	Luna 2	USSR	First hard landing, magnetic field
4.	4 Oct 1959	Luna 3	USSR	First photos of lunar far side
5.	23 Aug 1961	Ranger 1	USA	Attempted test flight
6.	18 Nov 1961	Ranger 2	USA	Attempted test flight
7.	26 Jan 1962	Ranger 3	USA	Missed the moon by 36,793 km
8.	23 Apr 1962	Ranger 4	USA	Crashed on the lunar far side
9.	18 Oct 1962	Ranger 5	USA	Missed the moon by 724 km
10.	2 Apr 1963	Luna 4	USSR	Missed the moon by 8,500 km
11.	30 Jan 1964	Ranger 6	USA	Hard landing, television failed
12.	29 July 1964	Ranger 7	USA	Hard landing, first close-up TV
13.	17 Feb 1965	Ranger 8	USA	Hard landing, close-up TV
14.	21 Mar 1965	Ranger 9	USA	Hard landing, close-up TV
15.	9 May 1965	Luna 5	USSR	Crashed on the moon
16.	8 June 1965	Luna 6	USSR	Missed the moon by 1,60,000 km
17.	18 July 1965	Zond 3	USSR	Photographed lunar far side
18.	4 Oct 1965	Luna 7	USSR	Crashed on the moon
19.	3 Dec 1965	Luna 8	USSR	Crashed on the moon
20.	31 Jan 1966	Luna 9	USSR	First soft landing and TV panorama
21.	31 Mar 1966	Luna 10	USSR	First lunar satellite, gamma-spectra, magnetic and gravity measurements
22.	30 May 1966	Surveyor 1	USA	Lander, on-surface TV, soil mechanics
23.	10 Aug 1966	Lunar Orb1	USA	TV imaging, radiation, micrometeorites
24.	24 Aug 1966	Luna 11	USSR	Orbiter, gamma and X-ray, measurements, gravity, micrometeorites

25.	22 Oct 1966	Luna 12	USSR	Orbiter, TV imaging
26.	6 Nov 1966	Lunar Orb 2	USA	TV imaging, radiation, micrometeorites
27.	21 Dec 1966	Luna 13	USSR	Lander, on-surface TV, soil mechanics
28.	5 Feb 1967	Lunar Orb 3	USA	TV imaging, radiation, micrometeorites
29.	17 Apr 1967	Surveyor 3	USA	Lander, on-surface TV, soil mechanics
30.	4 May 1967	Lunar Orb 4	USA	TV imaging, radiation, micrometeorites
31.	19 July 1967	Explorer 35	USA	Orbiter, Plasma, fields and particles
32.	1 Aug 1967	Lunar Orb 5	USA	TV imaging, radiation, micrometeorites
33	8 Sep 1967	Surveyor 5	USA	Lander, on-surface TV, first chemistry
34.	7 Nov 1967	Surveyor 6	USA	Lander, on-surface TV, chemistry
35.	7 Jan 1968	Surveyor 7	USA	Lander, on-surface TV, chemistry
36.	7 Apr 1968	Luna 14	USSR	Orbiter, gamma-spectra, magnetic measurements
37.	14 Sep 1968	Zond 5	USSR	First lunar flyby and earth return
38.	10 Nov 1968	Zond 6	USSR	Lunar flyby and earth return
39.	21 Dec 1968	Apollo 8	USA	First humans to orbit the moon
40.	18 May 1969	Apollo 10	USA	First docking in lunar orbit
41.	13 July 1969	Luna 15	USSR	Failed robotic sampler
42.	16 July 1969	Apollo 11	USA	First humans on the moon (20 July)
43.	6 Aug 1969	Zond 7	USSR	Lunar flyby and earth return
44.	14 Nov 1969	Apollo 12	USA	Human landing, Oceanus Procellarum
45.	11 Apr 1970	Apollo 13	USA	Aborted lunar landing
46.	12 Sep 1970	Luna 16	USSR	First robotic sample return, Mare Fecunditatis
47.	20 Oct 1970	Zond 8	USSR	Lunar flyby and earth return

No.	Launch date	Mission	Country	Accomplishment
48.	10 Nov 1970	Luna 17	USSR	First robotic rover Lunokhod 1, Mare Imbrium
49.	31 Jan 1971	Apollo 14	USA	Human landing, Fra Mauro
50.	26 July 1971	Apollo 15	USA	Human landing, Hadley-Apennine
51.	2 Sep 1971	Luna 18	USSR	Failed robotic sampler
52.	28 Sep 1971	Luna 19	USSR	Orbiter, lunar gravity, T V, micrometeorites
53.	14 Feb 1972	Luna 20	USSR	Robotic sample return, Apollonius
54.	16 Apr 1972	Apollo 16	USA	Human landing, Descartes
55.	7 Dec 1972	Apollo 17	USA	Human landing, first geologist on the moon, Taurus-Littrow
56.	8 Jan 1973	Luna 21	USSR	Robotic rover Lunokhod 2, Le Monier
57.	10 Jun 1973	Explorer 49	USA	Non-lunar radio astronomy from lunar orbit
58.	29 May 1974	Luna 22	USSR	Orbiter, lunar gravity, T V, micrometeorites
59.	28 Oct 1974	Luna 23	USSR	Failed robotic sampler
60.	9 Aug 1976	Luna 24	USSR	Robotic sampler return, Mare Crisium
61.	24 Jan 1990	Hiten	Japan	Flyby and orbiter, technological experiments
62.	25 Jan 1994	Clementine	USA	Orbiter, imaging lunar surface in U V, VIS, IR, laser altimetry
63.	6 Jan 1998	Lunar Prospector	USA	Gamma-neutron-alpha spectrometry, magnetometry, gravity
64.	27 Sep, 2003	SMART-1	ESA	Solar electric propulsion, near IR and X-ray Spectrometer
65.	14 Sep, 2007	Selene/ Kaguya	Japan	Mapping of lunar topography, surface composition and magnetic field
66.	24 Oct, 2007	Chang'-e	China	Explore lunar surface

				environment, topography, geological structures.
67.	2008	Chandrayaan-I	India	High resolution chemical, mineralogical and photo geological mapping of lunar surface in visible, near IR, low and high energy X-rays.
68.	2008	Lunar Reconnaissance Orbiter	USA	Obtain data to facilitate returning men safely to the moon

4

Beyond the Stratosphere: India's Space Programme

It is an inherent obligation of a great country like India, with its traditions of scholarship and original thinking and its great cultural heritage, to participate fully in the march of science, which is probably mankind's greatest enterprise today.

– The Scientific Policy Resolution, 1958

Below the stratosphere, in the verdant, remote islands of Lakshadweep in the Arabian Sea, frail 60-year-old Ayshomma was admitted to the Agatti community health centre, complaining of acute breathlessness. The government doctor on duty, Mubarak Baque, wanted a second opinion on her condition. With his patient unfit to travel by ship and unable to afford an air ticket, Baque resorted to the latest tool in his kit – telemedicine.

Ayshomma was wheeled into a new squeaky-clean air-conditioned room, empty except for a high-end computer. Hooking up through an INSAT satellite connection, Baque was soon chatting with Dr Krishna Kumar, a specialist at the Amrita Institute of Medical Sciences in Kochi, Kerala. In no time, a detailed consultation was underway, with Kumar being able to see Baque and Ayshomma. He viewed the lung X-ray and blood reports that had been digitally transmitted to him, discussed pros and cons with Baque and offered expert advice. At the end of a long video conference, Kumar confirmed acute bronchitis, allaying Baque's fears of a more serious condition. Ayshomma was put on a new course of medicine and everything suddenly started to seem under control for Baque.

Ayshomma is a clear beneficiary of advances in telemedicine. Over the past few decades, a large network has been put in place for which the uplink and downlink hardware has been provided by ISRO. Both private and government hospitals are slowly linking up with each other to offer super-speciality services in remote areas. The government hospital in Leh, situated in the high-altitude cold desert of Ladakh, is an institution where many lives have been saved through telemedicine networking with specialists at the Post Graduate Institute of Medical Research and Education, Chandigarh and the All India Institute of Medical Sciences, New Delhi. Other participating centres include Aizawl in Mizoram and Port Blair in the Andaman and Nicobar Islands.

Much like Ayshomma's story is that of B.R. Blah, a middle-aged government employee living in Cherrapunjee, Meghalaya, who used the digital network at the local Community Information Centre (CIC) to link up with the Christian Medical College, Vellore (over 3,000 km from his home) to seek a second opinion on his throat cancer.

Technology was always meant to do this – fill the gap, act as the great leveller between the haves and the have-nots. And telemedicine, a space technology application, is one area in which its impact is evident.

A functional telemedicine unit consists of a telemedicine-consulting centre like the one at Agatti Island and a suitable telemedicine speciality centre like the one at the Amrita Institute in Kochi. The consulting centre where the patient is housed must have necessary equipment for basic blood and urine analysis, besides a good scanner to convert the X-ray film into digital files, and a streaming electro-cardiogram unit (an ECG machine that can transmit digital signals across the network). High-end diagnostic systems can also be appended to the basic unit, but the heart and soul of a functional telemedicine unit is a functional audio and webcam apparatus. On the other side of the digital tunnel is the specialist centre, where there must be equipment that can receive such data. An ordinary personal computer or a television screen with necessary broadcasting equipment works fine. A satellite link provides connectivity between the two centres and a specially developed software programme can ease and streamline the process of communication.

The real power and reach of telemedicine is still to be realized in India; it is hoped that it will really start to make a difference to people's health, especially to those who really need such support. The private sector, as always, is confident this can be done: the Apollo group calls it 'healthcare without boundaries'. The private sector also views telemedicine as a tool to provide high-end but affordable healthcare facilities for people living in remote areas – like a virtual doctor on call. Telemedicine may also fill the gap when disasters – natural or man-made – strike and there is a sudden need for a large number of doctors and nurses.

This view of the entire nearside surface of the moon was photographed by Apollo II. Courtesy: NASA

The moon's surface at close quarters. Courtesy: NASA

The phases of the moon. Courtesy: NASA

The full moon against a backdrop of the blackness of space and the blue of the earth's horizon. Courtesy: NASA

A view of the far side of the moon taken during the Apollo 16 mission. Courtesy: NASA

A scientist on a bicycle near the Polar Satellite Launch Vehicle (PSLV), Sriharikota, 1999.

Pallava Bagla near the PSLV, Sriharikota, 1999.

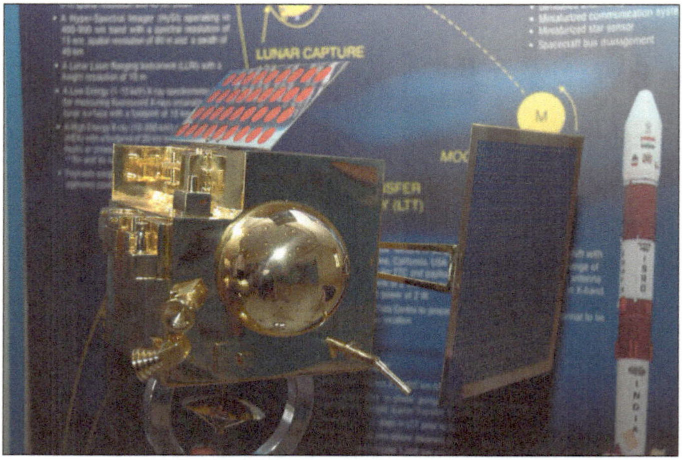

A scale model of Chandrayaan-1, Udaipur, 2004.

G. Madhavan Nair, chairman, ISRO, looking at the scale model of Chandrayaan-1, Udaipur, 2004.

The PSLV, Sriharikota, 2004.

PSLV successfully launches CARTOSAT-1 and HAMSAT, 5 May 2005.

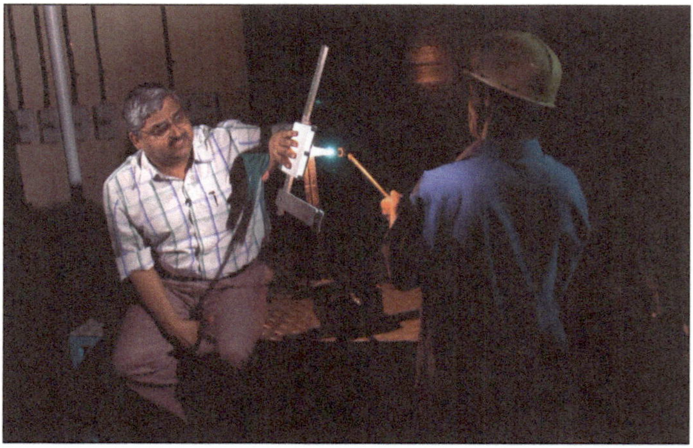

Pallava Bagla touches a ceramic tile developed by ISRO, which can withstand re-entry temperatures. The welder holds an oxy-acetylene flame on one side and Pallava can still touch the tile. Bangalore, January 2007.

The Satellite Recovery Experiment-1 capsule being recovered by the frogmen of the Coast Guard from the Bay of Bengal, in January 2007.

Sunita Williams and Rakesh Sharma at the International Astronautical Congress, Hyderabad, September 2007.

Dr Madan Lal with the Moon Impactor Probe being readied for the mission, Thiruvananthapuram, January 2007.

The Chandrayaan-1 satellite under fabrication at the ISRO Satellite Centre, Bangalore, February 2008.

Communication, connectivity and computers are surely the big Cs of change, and positive change at that. A little known success story is from the rather backward Kalahandi region of Orissa. In Silva, a small village in the region, on a potholed road lined with skinny bullocks is a small dusty room with a computer. The old IBM Pentium I machine, installed in 2000, opened up a whole new world for the villagers, allowing better tabs on the weather and hotlines to experts to seek solutions for problems with their cattle or crops.

Beyond these down-to-earth applications lies another high tech avatar of space commerce. At the ISRO headquarters in Bangalore, the Antrix Corporation, incorporated in September 1992 as the commercial arm of the Department of Space (DoS), sells classy remote-sensing pictures, robotic rides on space launchers and data products. It can also supply satellite systems and subsystems. The corporation offers training and consultancy services as well as ground systems and mission support.

Antrix has a clear-cut objective: to promote products and services of the Indian space programme and enhance their commercial value in the world. The commerce of science, research and development has always been a challenge for India, given the rigid socio-cultural framework of research for social good, and this has been true across disciplines and areas of enquiry. In diverse research laboratories and academic institutions of the country, getting scientists to think business and speak the 'pay-my-way' language has been difficult, to say the least. As a result, the transition of cutting-edge programmes from the bench to the bush is tough, and getting the boom into the business is even tougher.

The Antrix Corporation claims it is doing good business. The ISRO website says, 'The company has achieved steady and significant progress over the years. The activities of the company are predominantly export oriented and with more than 75% of the earnings being in foreign exchange, the company is earning valuable foreign exchange for the country.' Madhavan Nair says that according to the latest estimates, the sales turnover at Antrix is Rs 6,640 million.

Antrix is understood to be successful primarily because it has been built as a platform propped up by the Indian government, academia and industry. 'Its vision is to "emerge as a globally significant space company fully utilising the strengths of ISRO and other Indian entities in the field of space".'[1] Today, Antrix has the capability of providing end-to-end programmes for national needs, across the varied dimensions of application of space technology.

Is this then the world's window to India's success with its space programme? Is Antrix and the business it transacts of a level that is significant enough to call the Indian space programme a commercially viable proposition? Would Sarabhai have been thrilled with Antrix? He was, after all, devoted to the practical dimensions of space science, and wanted to make a difference in the lives of ordinary people.

The answers to all these questions lie in the dramatic trajectory of India's space programme.

In the late 1950s, about a decade into independence, the nation was just getting into its stride. National development with an eye on global trends was a major preoccupation for the government and for civil society. It was at this time that the doors to a 'new cosmic era'[2] opened, with Sputnik-1 being sent into space by the enterprising Russians.

Half a century later, Sputnik-1 is remembered as the very first space satellite made by humankind and sent successfully into orbit, and also as humanity's 'victory over gravitation'.[3] Even as Sputnik-1 began its exploration of the earth's environment, the very next month saw Sputnik-2 in space, with Laika the dog on board. The signals were clear: the world's superpowers were getting down to serious business beyond earthly matters, and the Russians were making history. India's portals of power were buzzing, catalyzed by Prime Minister Jawaharlal Nehru's interest in science as a driver of development.

From the time of independence, the country's leadership was convinced that science and technology would be crucial in pushing India ahead as a free nation. A Scientific Policy Resolution (SPR) – crafted by Nehru and Homi Bhabha – was passed in parliament in early 1958 and it became the first written commitment by the Indian government to 'secure for the people of the country all the benefits that can accrue from the acquisition and application of scientific knowledge' by fostering science and research and encouraging the initiation of training programmes. It said: 'The Government of India has decided to pursue and accomplish these aims by offering good conditions of service to scientists and according them an honoured position, by associating scientists with the formulation of policies, and by taking such other measures as may be deemed necessary from time to time'.[4] And in all this, the space programme had a very special place. It goes without saying that the SPR has played an important role in shaping the country's science and technology sector.

According to the Indian government's official media outreach website (www.pib.nic.in), 'When the country gained Independence in 1947 its S&T capability was at its lowest. The last fifty years have seen the country ranking third in the world as far as S&T personnel are concerned. There has been a prolific growth in the number of S&T institutions. The country's achievements in atomic energy, space and other sophisticated areas are well known.' Apparently, India was also the first country in the world to adopt such a resolution.

Even as Russia remained preoccupied with its new-found place in space, and Yuri Gagarin took off on the world's first space flight in 1961, India got down to working out the nuts and bolts of what promised to be a very complex and resource-intensive scientific programme. The Department of Atomic Energy (DAE) was set up under the leadership of Bhabha who, as its first secretary, was charged with steering the early days of space research and defining initiatives for the peaceful use of outer space. By 1962, the dynamic young Sarabhai, just 43 then, took charge of a space research entity called the Indian National Committee for Space Research (INCOSPAR), under the guidance of Bhabha.

Sarabhai, a trained physicist, had worked with Nobel laureate C.V. Raman at the Indian Institute of Science in Bangalore, researching cosmic rays. A driven man, he was fortunate in having Bhabha's unstinting support. Bhabha in turn was a deep and abiding influence on Nehru as far as shaping India's scientific vision and programme was concerned. In short, a fortuitous combination of excellence, vision, blessings from the top and great timing – all of this converged in a village in the deep south of Kerala, called Thumba.

Thumba is a medicinal plant that grows in abundance in this coastal fishing area. *Leucas aspera,* as it is called in Latin, is a herb with white flowers. Traditional healers use a paste of the leaves to offer relief from the sting of a scorpion. But these days, Thumba is a word less associated with flora and more with rockets and engineers. Work began on the Thumba Equatorial Rocket Launching Station in 1963.

In the same year, a Nike Apache sounding rocket from the US took off from this little place which was close enough to the magnetic equator to be able to make the most of certain special characteristics of the atmosphere in this region. A sounding rocket is so named because 'to take a sounding' means to take a measurement, in nautical language. This rocket could take necessary atmospheric measurements and perform scientific experiments too. For this baby step into the world of space, India roped in three majors, the US, Russia and France – quite a feat for a young country with an even younger science and technology programme.

A lot has happened since then. In 1969, the Indian Space Research Organisation (ISRO) was created from the INCOSPAR programme under the DAE. The DoS and the Space Commission were created in June 1972, and ISRO continued its activities under this banner.

During the defining decade of the sixties, Indian space initiatives truly started to take shape – from nascent launcher development to the first moves to develop a launch site to the conceptualization of a framework of action for ISRO for the next decade.

Now, it has opened its newest dimension of lunar missions with Chandrayaan-1, with its 11 payloads, five of which are from India and six from the US, ESA and Bulgaria. This has meant a serious gearshift for the Indian space programme: for the first time, its focus is scientific research rather than technology and applications.

Chandrayaan-2, an Indo-Russian venture, is to follow soon. The coming decade may also see developments towards India's first manned mission to the moon. That said, the programme has always focused on developing space technologies that could support national responsibilities, such as all dimensions of communication technology, the monitoring of resources and their management.

Satellites in Orbit

India has constantly been challenged by the need to develop effective communication channels across the country's vast geographical expanse, to be able to piece together information about natural resources as accurately as possible. The Indian National Satellite programme (INSAT) and the Indian Remote Sensing (IRS) satellite system form the core of the Indian space programme, focusing on communication and the use of remote sensing as a frontline technology for images that can be put to a variety of uses.

While INSAT is focused on communications and the strengthening of linkages between people through technology, IRS is a network built for better natural resource management. India currently has seven IRS satellites in service, the largest family of remote sensing satellites in the world in the civilian domain. CARTOSAT-2, sent up using PSLV-C7 in January 2007, is contributing to surveys for water resources, urban planning and crop acreage among other things, by beaming back high-resolution panchromatic images, with a resolution power that allows for distinguishing between two points just a metre apart. Put simply, it could be said that individual soldiers marching in a file could be distinguished as distinct images. CARTOSAT-2 images have been coming in for some time now.

Meanwhile, INSAT-4CR was launched on board the GSLV-F04 in September 2007 and is expected to make a difference to television broadcasting through Direct-To-Home (DTH) services. From telecommunication to television broadcasting, disaster warning to distance education, meteorology to telemedicine, the INSAT system is making a difference.

EDUSAT, launched in 2004, is another significant addition to this large domestic communications system (considered one of the largest in the world): it interlinks about 30,000 classrooms with academic institutions, universities, even primary schools.

The first phase of satellite technology, in the 1970s, was experimental (Aryabhata, Bhaskara, Rohini and APPLE – Ariane Passenger Payload Experiment). Aryabhata – a technology satellite named after the famous mathematician of ancient India – was the first entirely Indian satellite and carried payloads for aeronomy and solar physics experiments. Bhaskara was a precursor to the Indian remote sensing satellite series. Rohini and APPLE were also experimental satellites, and APPLE laid the foundation for geo-stationary communication. This was a critical phase in the programme, proving as it did India's capability to make space-worthy satellites and enhance technical expertise to operationalize communications satellites.

ISRO was able to move into an operational phase in the 1980s, with a focus on social and developmental benefit to the nation, and services were rapidly introduced. This is when INSAT and IRS evolved and finally became pillars of the Indian space programme.

There have been various milestones within the satellite programme (see Table 3.1). In December 2005, India sent up

GSLV launch from Sriharikota, 2 September 2007

its heaviest and most powerful satellite to date, INSAT-4A, programmed for DTH television broadcasting services. India has a whopping eleven national communications satellites in orbit simultaneously, the highest for any country in the Asia-Pacific region. This gives it the distinction of having one of the largest national networks of operational satellites anywhere in the world, no mean achievement for a country still struggling to keep a third of its population above the poverty line. But then, as ISRO says, for every rupee spent on the space programme, the return has been two rupees.

It is interesting to see how the two programmes, INSAT and IRS, converge over their applications in the Village Resource Centres (VRCs) concept mooted by the DoS some years ago.

Tuned to the needs of rural communities, the VRCs (which house interactive Very Small Aperture Terminals, or VSATs) provide telemedicine and education services, adult education packages, and health and family welfare initiatives. Started as a joint initiative of ISRO and the M.S. Swaminathan Research Foundation in Chennai (MSSRF) in October 2004, it has grown to 300 VRCs while also gathering several other partners like panchayats. It now addresses rural needs for education, health, nutrition, weather forecasting, environmental issues and agriculture-related information. The VRC, which uses the INSAT-3A satellite, is a single-window delivery system for all kinds of services that require satellite technology.[5]

MSSRF has connected ten villages near Pondicherry in southern India through a hybrid wired and wireless network that facilitates voice and data transfer, enabling villagers to get information that can fulfil their needs. This network can make available weather forecasts, market prices of commodities, knowledge of and access to better farming practices and new seed varieties, and enhance job opportunity by disseminating employment advertisements. Overall, it has the potential to truly improve a society's quality of life. The project drew sustenance from the holistic philosophy of Dr M.S. Swaminathan – an integrated pro-poor, pro-women, pro-nature approach to development that advocates community ownership of technological tools, thus encouraging collective action for the spread of technology. The delivery of electronic knowledge to the less privileged is at the heart of this philosophy.

Looking back, one realizes that the images from India's IRS network have attracted international attention over the past several years. Today, the global marketing of these data products is a core activity at Antrix. A partnership with the US company GeoEye has led to a robust network of ground stations spread across the world that receive and/or process data from the IRS satellites. IRS hardware and software are also marketed. Unconfirmed reports suggest that America used high-quality images of Afghanistan obtained with the help of Indian remote sensing satellites to target Osama Bin Laden's camps after 9/11. The calibre and technique of the IRS network is well known, making it popular among users of such data and information.

'Satellites for society' has been the unofficial mantra of the Indian space programme ever since it began. The Experimental Satellite Communications Earth Station (ESCES) was set up by Sarabhai in the late sixties, INSAT was commissioned in 1983 and IRS in 1988 – all of these a distillation of his vision that applications would be critical to the success and sustenance of the Indian space programme. So taken was the Indian space community with this emergent need to prove application, that satellite development was encouraged almost independently of launch capability. But if there's a satellite up in space, can a rocket be far behind?

Launchers and Other Stories

Today, India offers launch services on a commercial basis. Not so long ago, Indian satellites were piggybacking on foreign rockets. With his characteristic sense of urgency, Sarabhai had set up the Space Science and Technology Centre (SSTC) at Thiruvananthapuram in 1966. This was his signal to the country and the world that India was ready to take its second leap in the space programme – developing indigenous satellite launch capability. Renamed the Vikram Sarabhai Space Centre (VSSC) after his death, the institution is engaged in several rocket and launch vehicle projects. There are wide-ranging research and development activities that focus on avionics, aeronautics, materials and mechanical engineering, solid propulsion and composites, propellants, polymers and chemicals, systems reliability and control software.

The ISRO launch vehicles – rockets capable of transporting loads as diverse as satellites, human beings, or even dogs (recall Laika, the dog that became the first mammal to orbit the earth) – are designed and developed here, from the Satellite Launch Vehicle-3 (SLV-3), and the Augmented Satellite Launch Vehicle (ASLV) to the PSLV and the Geosynchronous Satellite Launch Vehicle (GSLV).

After a maiden failure, the SLV-3 made it in 1980. If Aryabhata was a milestone for the experimental phase of satellite technology, SLV-3 was one in terms of launcher expertise. But very quickly, rocket scientists knew they would need to move beyond, given that the SLV-3 could carry payloads of only about 40 kg. The answer was the PSLV, and a decision was made to move towards this technology. The ASLV was a comfort bridge for taking such a giant technological leap, but it took two failed launches to get there. And the PSLV was soon around the corner – as a result, so was the ability to place application satellites in orbit.

Orbits are a critical piece in the space puzzle. Various orbits exist around the earth, and satellites can be placed depending on the application they are meant for and the technology available. Low earth orbits are less than 2,000 km in altitude.

Satellites in the 1,600-kg category can be launched into a 620-km polar sun-synchronous orbit using the PSLV class of rockets. Placing a satellite in this orbit means it will cross the equator at the same longitude and time each day; being in a polar orbit would imply that on each revolution, the satellite would pass above or nearly above both poles of the planetary body it is orbiting.

The geosynchronous orbit exists more than 36,000 km above the earth, and around the equator. A geosynchronous transfer orbit is the first stop before satellites can reach the geosynchronous orbit and it is elliptical in nature. The 2,000-kg category of satellites can be placed into a geosynchronous transfer orbit using India's larger rocket, the GSLV.

Today, Indian rockets are capable of launching multiple payloads into different orbits with precision. In April 2008, the PSLV-C9 set a world record by launching 10 satellites in one go, the CARTOSAT-2A, and nine smaller satellites. Success is significantly influenced by this one factor. Earlier it was the PSLV-C7 that demonstrated this four-in-one capability in January 2007 by carrying CARTOSAT-2, India's best mapping satellite till date, with two small foreign satellites, the Indonesian 56-kg satellite Lapan-Tubsat and the 6-kg Argentinian PEHUENSAT-1, and the first ever re-entry experiment by India through the 550-kg Satellite Recovery Experiment-1 (SRE-1).

The SRE-1 marks a significant milestone in the quest to recover and successfully bring back to earth an orbiting space capsule. This was a critical manoeuvre that must be perfected if ISRO wants to put an Indian in space using Indian rockets. The SRE-1 made a big splash in the Bay of Bengal 12 days after it was sent into orbit and was pulled out by frogmen of the Indian Coast Guard. The event also made news given how challenging it is to master re-entry technology; to the best of our knowledge no other country has successfully recovered an orbiting satellite in the very first attempt.

Then there was the PSLV-C8 launch in April 2007, carrying the Italian satellite AGILE into a low earth orbit. This was the first time a foreign buyer hired the entire vehicle for the launch of its satellite. For Antrix, the AGILE launch was another milestone in the commercialization of space products, as ISRO charged a commercial fee higher than the standard market rate for this maiden commercial launch.

Inking its second full-fledged international commercial contract, ISRO used the PSLV to launch the Israeli military satellite TecSAR into orbit on 21 January 2008 from Sriharikota. TecSAR, weighing 300 kg, is what is informally called a spy satellite, with the capability of capturing images both in the day and the night, as well as in cloudy conditions. There was unprecedented secrecy around the event although TecSAR was launched under a commercial agreement between Israel Aerospace Industries and Antrix. Unlike on other occasions, there were no exhaustive announcements before the launch.

Foreign clients form two-thirds of Antrix's business. ISRO charged about Rs 44 crore (approximately US $11 million) to send AGILE into orbit. While the charge for sending TecSAR into space was not revealed, ISRO confirms it was higher than the market rate of Rs 8 lakh or US $20,000 per kg. Earlier, in May 1999, a 100-kg Korean satellite KITSAT-3 and a 50-kg German satellite were sent into space aboard the PSLV-C2, for which too, ISRO claims to have charged commercial rates.

In September 2007, using the GSLV-F04, Indian space scientists managed to successfully place in orbit INSAT-4CR – a replacement for the satellite that had been shattered in the only failed mission of the GSLV. This spectacular failure of July 2006 was attributed to a malfunction in a strap-on motor on the GSLV rocket. The GSLV-F04 success was notched up hot on the heels of that disaster.

The PSLV has successully flown 12 times consecutively till now and it was only its first flight that failed. In April 2008 this tried-and-tested rocket flew successfully for the thirteenth time, taking ISRO's score to 26 launches. In fact, the April 2008 launch was a world record for ISRO: it took 10 satellites up into orbit at one go.

How did India come to this point of launch capability? In the early days, the SSTC was a 'technological melting pot',[6] with groups of engineers and scientists working on different dimensions of rocket science. Whether it was to do with developing solid propellants or sounding rocket designs or special plastics that are reinforced with fibres, teams were busy from day one and, as always, Sarabhai's energy and enthusiasm were infectious. Many feel that his sudden death in his sleep on 30 December 1971, in Kerala, left a major void that has been hard to fill.

The Indian space programme has certainly grown against great odds. Severe technology denial regimes ensured that state-of-the-art technologies that were available to other countries off the shelf, were denied to India. Then, crippling sanctions were imposed on the country soon after it demonstrated its capability of exploding a nuclear device in 1974 at Pokharan. The US spearheaded these sanctions and their adverse impact was most evident when in the early 1990s, succumbing to American pressure, Russia declined to transfer to India the technology to make cryogenic engines. These sophisticated rocket motors that use liquid oxygen and liquid hydrogen for propulsion are essential for lofting large communications satellites. Now that the Indian cryogenic engine has been assembled, the next slotted flight of the GSLV in 2008 in all probability should use the indigenously developed cryogenic stage. India is the sixth country to have mastered this technology.

But several ISRO labs still languish on the dreaded 'entity list' – a roster of institutions that require special export licences to be issued by the American bureaucracy – even for the export of simple computer chips found in ordinary laser printers and washing machines.

The Sarabhai Factor, Systems and Structures

Former chairman of ISRO Kasturirangan wrote in the Indian science monthly *Resonance*:

> It is always the first few steps that are important in the life of an organization. Those steps are crucial and need to be taken under the gentle care of a humane and loving leader, who can reassure the organization and reinvigorate it at every faltering step, admire every right step and encourage and aid in taking larger steps ahead. If ISRO, the organization that implements the space program in India, has made an indelible mark in the development and application of space technology, it is because of the crucial leadership that was provided at its infancy by Vikram Sarabhai, truly, the father of the Indian space program.

Sarabhai provided enormous impetus to the nascent Indian space programme. He ensured that his team got world-class training, and helped them stay abreast of global developments, in the difficult early days when the country could not even feed itself and was living ship to mouth. So many decades later, shades of that early spark – the will to strengthen the skill and capacity of Indians to confront the challenges presented by cutting-edge space science and become self-reliant – are manifesting themselves near Thiruvananthapuram, where the campus of the Indian Institute of Space Science and Technology (IIST) is coming up. Set up in September 2007 to spur young people to take up specialized undergraduate and postgraduate courses with a particular focus on space-related subjects, IIST has already enrolled its first batch of 140 students. The core mission here is to offer 'high quality education in space science and technology to meet the growing demands of the Indian space programme'.

The space programme needs skilled young people, but given the range of career options available to Indians students today, it is no small challenge to draw them to the field. That said, young people with an inclination towards advanced higher studies in the fields of science and technology have often felt the need to delve into the applied dimensions of science. That is exactly what IIST plans to offer through integrated undergraduate and masters programmes in which students can study space technology, avionics, aerospace engineering, atmospheric sciences, material sciences and other disciplines. Its aim is to integrate education and applied research using state-of-the-art technology, and develop a high-calibre teaching faculty.

Dr B.N. Suresh, director of the nascent IIST (and former director, VSSC, Thiruvananthapuram, says, 'Today ISRO has attained an end to end capability of managing the entire space services needed for the country. A question that is often posed is: how could India, a developing country, reach this level of maturity in sophisticated technology like space, and that too, starting from scratch?'[7]

If only Sarabhai were around to answer this question. When he took charge of the space programme, he knew he had a towering challenge before him. In the early sixties of the previous century, the requisite skills and expertise were rare among Indians, and the infrastructure woefully inadequate. In fact, in the early days, the picturesque St Mary Magdalene church in Thumba was turned into a space lab – the first such laboratory in the country – since there was no other building for the many staffers to function from. As Dr Kalam is known to have said:

> Designing and developing the first launch vehicle for our country was a magnificent effort of over 10,000 scientists, engineers, staff members, technicians, industrialists, academicians and many others; we didn't have modern tools like CAD packages, push button phones, computers. The project team [for SLV3, India's first launch vehicle deployed successfully in 1980] was housed in tin roof sheds![8]

Early architects like Sarabhai also knew that they were a long way from convincing their critics that all of this was being done for the benefit of society. What worked then for the space programme, in a nation where all gates to development had to be opened at the same time? There is no doubt that the space programme was a beneficiary of extraordinary political commitment that cut across party lines. Even today, the prime minister of India holds the portfolio of the DoS.

All said, perhaps the most critical ingredient in the winning formula has been simple and straightforward – openness and transparency translating into a strong public relations and outreach programme, an ability to accept failure with calm professionalism and great élan and the scientific and technological skill to address the causes of failure each time, every time. ISRO is well known for its prompt and factual press notes and briefings – whatever the outcome of its activities.

More importantly, the programme benefited from a unique brand of leadership that certainly deserves admiration, and hopefully, emulation in other large national programmes and institutions. When the GSLV launch failed in 2006, it was Madhavan Nair who led the press conference, taking all questions and trying to supply convincing answers. Years ago, when the SLV-3 failed, it was a black day for the community of space scientists and engineers in India. Dr Kalam had been the leader of the project, but Prof. Satish Dhawan, chairman of ISRO at the time, surprised everybody on the team by saying he would address the press conference. The very next year, when the launch was successful, Dhawan invited Kalam to host the press conference. When Kalam spoke at public forums years later, as the president of India, this was one of his favourite stories – he would use it as an example of strong, effective leadership.

ISRO is also known to have delivered on target, time and again. As Nair said:

> The fact that Indian technology has been marketed globally in the form of IRS images, launch services and many sub-systems, and assemblies for spacecrafts, shows the maturity and competitiveness of ISRO. The entire space activities in the country have been

> managed in a cost-effective manner and the reliability of the space systems has been proven through a number of missions.[9]

Benefit to people is a core value for the Indian space programme: 'the application of advanced technologies to the problems of man and society', as Sarabhai was so fond of saying. Sarabhai's wife and well-known dancer Mrinalini Sarabhai, once wrote in *Resonance*: 'He saw tremendous potential of modern science to help in the process of development. He drew up plans to take modern education to the remotest villages by using satellite television.' His interest in education resulted in his establishing the Community Science Centre, an institution where new ideas in science education for children could be tried out. He often said that on retirement he would like to spend time with young children and talk to them about science.

Kasturirangan wrote in 2001:

> Three decades later, with his dream fulfilled by and large, we have unambiguously realized that Sarabhai had a rare vision. Today, the nation is reaping the benefits of our space systems like INSAT and IRS in various areas like telecommunication, TV broadcasting including innovative use of the media for developmental communication and education, meteorology including disaster warning, resources survey and their management. But what is important to note is that, in the early sixties, space was still a new technology and many nations including developed ones considered it esoteric, expensive and not applicable to developing countries. But here was a scientist in a developing country visualizing the

> application of a new technology, seeing its potential for connectivity and ultimately creating the right tenor for realizing his vision of India effectively, making use of it. It was Sarabhai who conceptualized the Satellite Instructional Television Experiment (SITE) which was one of the largest sociological experiments conducted anywhere in the world involving educational television broadcasting via the US satellite, ATS-6, to rural masses spread over six states and covering 2,300 villages. He was determined to prove the utility of a space system for grassroots level application for a large country like India. He proved the efficacy of space-based remote sensing through organization of airborne experiments for detecting coconut wilt disease in Kerala in the sixties. It was he who proposed the first Indian experimental satellite, later named Aryabhata, to acquire first hand experience of designing, building and operating a satellite system.[10]

In India today, it does seem as though technology has raced a few paces ahead – maybe several – of the social and developmental fabric as well as the infrastructure essential for effective absorption of all that is on offer from such sophisticated technology. For instance, if EDUSAT is a communications satellite that can connect classrooms (which it is already doing), its real value and lasting impact will lie in the vibrancy of that network, its optimal function and sustained quality over time. Recent reports regarding the contribution that EDUSAT is making to 'real' education have asked the all-important question – as the project approaches mid-life (EDUSAT was launched in 2004), has it really given quality education, bridged the gap, and reached where no education could otherwise reach? The answer is that EDUSAT is sadly under-utilized.[11]

The bottlenecks are all the obvious ones – classrooms with EDUSAT link networks are often empty, the modules seem extraneous to primary curriculum and routine, and the integration that should accompany the introduction of high-end technology was perhaps not planned. As is often the case, the objectives were laudable and the heart in the right place, but the lack of an inclusive approach that took account of the people on the ground (such as teachers and education experts) resulted in serious lacunae in the viability of the technology. Madhavan Nair clarifies that a powerful communications satellite like EDUSAT is only a tool, after all, and its effectiveness lies in the hands of the educationist and the content generator.

> The concerned state education departments have to take the initiative. I think at least half of the states which have been provided these facilities are doing well. Others, again, you know it depends on their individual interests. We enable them, but unless there is a person who can drive that, sometimes you know it does not catch up. There is still scope for improvement.

The question mark over the ability of the system to absorb high-end technology is not new for ISRO. More than a decade ago, the remote sensing programme had a lot of explaining to do – a great deal of investment had gone in and high-calibre data and images were beginning to flow back, but users were complaining of cumbersome access and delay in receiving products.[12] Hurdles included archaic government policies that resulted in data share issues with the general public. Fortunately, ISRO as an organization is good at learning and moving with the times.

Keeping the Faith, Making a Difference

Many Indians living and working in the present day have clear recall – after all, it was just two decades ago and a bit – of the time when phone calls were expensive, a telephone instrument was a luxury product and even the ability to communicate over the phone was reserved for the privileged few. Indian satellites in space changed all that, as ISRO got into the business of connecting India's billion-plus people. These satellites ushered in a telecom revolution in the 1980s. Subscriber Telephone Dialling or STD became the buzzword, breaking language barriers and shrinking the vast distances across India.

The connections that Sarabhai had dreamt of were happening, and how! Known for his obsessive focus on making sense out of space science for the ordinary Indian, Sarabhai, speaking in Bangalore in 1966, had said:

> Physical and financial inputs are of course essential for their [the people of India] economic and social uplift, but surely an essential ingredient of our success is the ability to communicate with the widely dispersed population in the region for improving agricultural productivity and permitting programmes of population control. Space research also confers an invaluable (but sometimes intangible) advantage through the spread of advanced technologies which are related to economic

development and security. These technologies should materially assist developing countries such as India to leap-frog from their present status. [13]

The Indian space programme seems to have kept faith with its much revered founder. In the past decade, the penetration of the internet in rural India has been spurred on by satellites in space. Remote parts of India are getting connected to the mainstream and the benefits of the worldwide web can be seen in places falling off the map – and there are many such places in India. For instance, Anini is a name that may evoke images of a Hawaiian beach for some. Actually, it is one of India's most far-flung district headquarters – it takes a minimum of 96 hours to travel from New Delhi to Anini, in the Dibang valley district of the north-eastern state of Arunachal Pradesh, strategically perched on the China border. There is no all-weather road connecting Anini with the outside world, which is why it is called the 'hidden paradise'.

There cannot be too many places in the world where the information super-highway has appeared before a tar road – as it did here, using ISRO's INSAT satellites. Anini today houses a CIC – part of the 487 CICs (all linked using ISRO's INSAT satellites) set up all over the north-east by the Ministry of Information Technology (MIT) of the Government of India, the National Informatics Center (NIC) and the respective state governments of the north-east, in an effort to bridge the digital divide in this secluded corner of India. Basic services to be provided by CICs include internet access and e-mail, printing, data entry and word processing and training for the local populace; in addition, several citizen-centric or government-to-citizen services are available.

This was at the beginning of the new millennium, in 2002. Amazingly, the six Pentium computers and other peripherals provided to the Anini CIC arrived – with majesty and grace – on elephant back. There was no choice, actually, since there was no connectivity by road. An aerial public distribution system has always been the saving grace for this remote place, with Indian Air Force helicopters air-dropping rice, wheat, cooking oil, and even people. The nearest road-head is Roeing but the Anini–Roeing road, built by the Border Roads Organisation, is often blocked by landslides.

Rajesh Thacho, a political leader based in Anini, acknowledges that the track remains largely un-serviceable, traversing as it does acres of lush rainforest. So the computer centre was obviously big news, especially for the Idu Mishmi tribe for whom this place is home. It was not in any way an indulgence, more a necessity to bridge the communication gap that separated the region from the rest of India. Especially as electric power had already taken care of – a 250-kw micro-hydel project had been commissioned by the state government.

At the Mohanbari air force base, Ashish Mokashi, chief operating officer for the helicopter unit servicing the region, reflected, 'Young children see a helicopter and an airplane well before they get to see a bicycle or a car.'

Another CIC came up in the north-east, at Cherrapunjee, better known for the huge amounts of rain it receives – roughly 12 metres a year! Soon after being set up, the CIC showed results: the Block headquarters were computerized. Apparently, internet connectivity was not a problem, even during peak monsoon conditions.

Making headway with disaster management systems has always been a core priority for the Indian space programme.

Idu Mishmi tribals at the CIC computer centre, Anini, Arunachal Pradesh, August 2002

It used a synergistic approach, drawing on the strengths of both remote sensing and communications satellites, along with ground networks and capability. The country's tsunami warning system uses this network and so does the mechanism for cyclone warning.

ISRO has installed a whole series of cyclone warning stations on the east coast of India. About a decade ago, the India Meteorological Department (IMD) collaborated with ISRO to develop a cyclone early warning system that has since helped save many lives. It is an elegant concept, with a seemingly simple system of operation. Information about impending cyclones is transmitted by IMD to ISRO satellites, which in turn relay the message to automatic hooters installed in remote coastal villages. A shrill blast goes off, signalling that a cyclone of high intensity is about to strike. Early warnings do make a difference, but finding the automated method limiting, efforts were made to incorporate a human element into the process, with local residents being used to spread the word. This was experimented with in the cyclone-prone belt of the east coast.

What has always stood ISRO in good stead, and taken the institution from strength to strength, is its record with transparency and openness of communication. Whether a successful project or not, whether a copy-book launch or one gone wrong for whatever reason, the approach is straight and steady – prompt press releases and outreach, detailed fact-finding exercises and failure analysis reports, and open sharing of as much detail as can be permitted in this strategic sector. It is no exaggeration to say that other national programmes would do well to take a few pages out of ISRO's book.

There is little doubt that the Indian space programme is robust and progressive. A few years ago, in 2002, Prof. Narendra Bhandari, then a senior scientist at the planetary science and exploration programme at the Physical Research Laboratory, Ahmedabad is known to have said that the PSLV 'is capable of undertaking missions for direct exploration of Mars and Venus'. Bhandari was not alone in dreaming big. Dr Kalam, too, before becoming the president of India, had expressed his unambiguous opinion that 'India should concentrate on a mission to Mars rather than the moon as it would be more useful scientifically'. He had added that 'Mars has good potential for space industrialization'. So, even as Chandrayaan-1 becomes a milestone, Mars may not be far behind.

Table 3.1: Highlights and Milestones: The Indian Space Programme (1962 to 2008)

(*Source: www.isro.gov.in*)

2008	ISRO's Polar Satellite Launch Vehicle, PSLV–C9 successfully launches two Indian satellites, the larger CARTOSAT-2A, and eight small satellites from other countries (28 April 2008). PSLV–C10 successfully launched Israeli satellite, TECsar (21 January 2008).
2007	PSLV–C8, successfully launched Italian astronomical satellite, AGILE from Sriharikota (23 April 2007). Successful launch of INSAT–4B by Ariane-5 from Kourou, French Guiana (12 March 2007). Successful recovery of Satellite Recovery Experiment SRE-1 after manoeuvring it to re-enter the earth's atmosphere and descend over the Bay of Bengal about 140 km east of Sriharikota (22 January 2007) PSLV–C7 successfully launches four satellites – India's CARTOSAT–2 and SRE-1 and and Indonesia's LAPAN–TUBSAT and Argentina's PEHUENSAT–1 (10 January 2007).
2006	Second operational flight of GSLV (GSLV–F02) from SDSC SHAR with INSAT–4C on board. Satellite could not be placed in orbit (10 July 2006).
2005	Successful launch of INSAT–4A by Ariane from Kourou, French Guiana, (22 December 2005). ISRO's PSLV–C6, successfully launched CARTOSAT–1 and HAMSAT satellites from Sriharikota (5 May 2005).
2004	The first operational flight of GSLV (GSLV–F01) successfully launched EDUSAT from SDSC SHAR, Sriharikota (20 September 2004).
2003	PSLV–C5, successfully launched RESOURCESAT–1 (IRS-P6) satellite from Sriharikota (17 October 2003). Successful launch of INSAT–3E by Ariane from Kourou, French Guiana (28 September 2003). The second developmental launch of GSLV–D2 with GSAT–2 on board from Sriharikota (8 May 2003). Successful launch of INSAT–3A by Ariane from Kourou, French Guiana (10 April 2003).

2002	PSLV–C4, successfully launched Kalpana-1 satellite from Sriharikota (12 September 2002). Successful launch of INSAT–3C by Ariane from Kourou, French Guiana (24 January 2002).
2001	PSLV-C3, successfully launched three satellites – Technology Experiment Satellite (TES) of ISRO, BIRD of Germany and PROBA of Belgium – into their intended orbits (22 October 2001). The first developmental launch of GSLV–D1 with GSAT–1 on board from Sriharikota (18 April 2001).
2000	INSAT–3B, the first satellite in the third generation INSAT–3 series, launched by Ariane from Kourou, French Guiana, (22 March 2000).
1999	Indian Remote Sensing Satellite, IRS–P4 (OCEANSAT), launched by PSLV–C2 along with Korean KITSAT–3 and German DLR–TUBSAT from Sriharikota (26 May 1999). INSAT–2E, the last satellite in the multipurpose INSAT–2 series, launched by Ariane from Kourou, French Guiana (3 April 1999).
1998	INSAT system capacity augmented with the readiness of INSAT–2DT acquired from ARABSAT (January 1998).
1997	INSAT–2D, fourth satellite in the INSAT series, launched (4 June 1997). Becomes inoperable on October 4, 1997. (An in-orbit satellite, ARABSAT–1C, later renamed INSAT–2DT, was acquired in November 1997 to partly augment the INSAT system.) First operational launch of PSLV with IRS–1D on board (29 September 1997). Satellite placed in orbit.
1996	Third developmental launch of PSLV with IRS–P3 on board (21 March 1996). Satellite placed in polar sun synchronous orbit.
1995	Launch of third operational Indian Remote Sensing Satellite, IRS–1C (28 December 1995). INSAT–2C, the third satellite in the INSAT–2 series, launched (7 December 1995).
1994	Second developmental launch of PSLV with IRS–P2 on

	board (15 October 1994). Satellite successfully placed in polar sun synchronous orbit.
	Fourth developmental launch of ASLV with SROSS–C2 on board (4 May 1994). Satellite placed in orbit.
1993	First developmental launch of PSLV with IRS–1E on board (20 September 1993). Satellite could not be placed in orbit.
	INSAT–2B, the second satellite in the INSAT–2 series, launched (23 July 1993).
1992	INSAT–2A, the first satellite of the indigenously-built second-generation INSAT series, launched (10 July 1992)
	Third developmental launch of ASLV with SROSS–C on board (20 May 1992). Satellite placed in orbit.
1991	Second operational Remote Sensing satellite, IRS–1B, launched (29 August 1991).
1990	INSAT-1D launched (12 June 1990).
1988	INSAT-1C launched (21 July 1988). Abandoned in November 1989.
	Second developmental launch of ASLV with SROSS–2 on board (13 July 1988). Satellite could not be placed in orbit.
	Launch of first operational Indian Remote Sensing Satellite, IRS–1A (17 March 1988).
1987	First developmental launch of ASLV with SROSS–1 satellite on board (24 March 1987). Satellite could not be placed in orbit.
1984	Indo-Soviet manned space mission (April 1984).
1983	INSAT–1B launched (30 August 1983).
	Second developmental launch of SLV–3. RS–D2 placed in orbit (17 April 1983).
1982	INSAT–1A launched (10 April 1982).
	Deactivated on 6 September 1982.
1981	Bhaskara–II launched (20 November 1981).
	APPLE, an experimental geo-stationary communication satellite successfully launched (19 June 1981).
	RS–D1 placed in orbit (31 May 1981) by the first developmental launch of SLV–3.

1980	Second Experimental launch of SLV–3, Rohini satellite successfully placed in orbit (18 July 1980).
1979	First Experimental launch of SLV–3 with Rohini Technology Payload on board (10 August 1979). Satellite could not be placed in orbit. Bhaskara–I, an experimental satellite for earth observations, launched (7 June 1979).
1977-79	Satellite Telecommunications Experiment Project (STEP) carried out.
1975-1976	Satellite Instructional Television Experiment (SITE) conducted.
1975	First Indian Satellite, Aryabhata, launched (19 April 1975). ISRO becomes government organization (1 April 1975).
1972-1976	Airborne remote sensing experiments.
1972	Space Commission and Department of Space set up (1 June 1972). ISRO brought under DoS.
1969	Indian Space Research Organization (ISRO) formed under Department of Atomic Energy (15 August 1969).
1968	TERLS dedicated to the United Nations (2 February 1968).
1967	Experimental Satellite Telecommunication Earth Station set up at Ahmedabad.
1965	Space Science & Technology Centre (SSTC) established in Thumba.
1963	First sounding rocket launched from TERLS. (21 November 1963).
1962	Indian National Committee for Space Research (INCOSPAR) formed by the Department of Atomic Energy and work on establishing Thumba Equatorial Rocket Launching Station (TERLS) started.

5

Chandrayaan: India's Shot at the Moon

Twenty years from now when space travel is likely to become mundane like airline travel today, we don't want to be buying travel tickets on other people's space vehicles.

– Dr G. Madhavan Nair, Chairman, ISRO

The Plan, the Mission, and the Goals

India's inclination for pushing the envelope on world-class science and technology – especially in the area of space science – is one of the reasons Chandrayaan-1 was born in the first place. This is reflected in the formally stated objectives of the mission: to 'expand scientific knowledge about the origin and evolution of the moon, upgrade India's technological capabilities and provide challenging opportunities to young scientists working in planetary sciences'.[1]

It makes sense to master the techniques central to planetary exploration because in the not-so-distant future, humans will be looking towards space and planets other than the earth to support the world's energy needs. So why miss the bus – or the rocket, in this case? But this has meant a historic transition for India: after more than three decades of being a space-faring nation that relentlessly focused on looking down towards the earth from space, the country is training its sights on outer space as a place to explore.

On 11 May 1999, marking the first anniversary of Pokharan II, the First Technology Day was organized in New Delhi. The special day was dedicated to the nation by Prime Minister Atal Behari Vajpayee at the behest of Union Minister for Science and Technology Prof. Murli Manohar Joshi. Delivering the first lecture, Dr Kasturirangan, chairman of ISRO, spoke about the capabilities of India's workhorse rocket, the PSLV. Everyone was in a celebratory mood and all attention was focused on trying to fathom the immense political and international fallout of India having declared itself a nuclear weapons state a year earlier. Nobody had a clue that another giant technological step for the country was to be announced.

It was a buzzing, busy day, with scores of science-related events taking place and things generally running behind schedule. Kasturirangan's talk was also delayed and finally began rather late in the evening, quite fortuitously, when the moon was shining in all its glory. Towards the end of his hour-long talk, Kasturirangan, without fanfare and with a serenity that only he could display while making such a historic announcement, gently slipped in a seemingly unremarkable comment – that the PSLV could undertake a mission to the moon.[2] Few recognized that India's space science community meant business, serious business. Elaborating on the plan, Kasturirangan said that 'working out the mission's objectives and payload could take time and if all goes well, it could be a reality by 2008.'

It took ISRO four years to get an endorsement from the government. The news was made public by Prime Minister Vajpayee on Independence Day, 2003, in his address to the nation from the Red Fort in New Delhi: 'India is ready to fly high in the field of science...I am happy to announce that by 2008 India will send a mission to the moon.' It was to be called 'Chandrayaan Pratham' or the first journey to the moon. Also translated as Mooncraft-1, the name clearly indicated that India was planning additional missions.[3] Chandrayaan-1 was to be the country's first probe to venture beyond the orbit and to map the moon's surface in detail.

It was in November 1963 that India sent off its first sounding rocket into the earth's atmosphere. This became the first key activity of India's space programme. For Indians born in the 1960s (the generation that today forms the bulk of the country's productive workforce), the transformation is near magical: what began as a few lines in a dry school science textbook is today an ambitious programme with global ramifications.

Going back to May 1999, Kasturirangan's announcement and the prominent headline in the *Indian Express* the following day[4] were followed by an editorial blitz with commentators questioning why India wanted to repeat a feat that had been achieved decades ago by the US. They also asked whether the moon had lost its sheen in the ensuing years because if it was so attractive, why had the US abandoned all exploratory activities? Of course, the same commentators had to rethink their position when US President George W. Bush announced in 2004 America's renewed interest in going to the moon. It is also worth noting that India's announcement in 1999 was ahead of China's announcement of a mission to the moon in 2004. Many former critics of ISRO's plan now started to say that going to the moon was a very good idea since the US was charting a new plan to create a permanent base on the moon in the years beyond 2020.

Just about two weeks after Kasturirangan's talk alluding to India's designs on the moon, Vajpayee was at Sriharikota in Andhra Pradesh, on 26 May 1999. He was there to witness the launch of OCEANSAT, the Indian Remote Sensing Satellite IRS-P4, launched by PSLV-C2 along with two micro satellites, the Korean KITSAT-3 and the German DLR-TUBSAT. The country's top leadership does not grace the send-off of every satellite into space, so this was clearly special. It was also the first time that other nations had booked tickets on an Indian trip to space, a sign that the world was beginning to take serious note of India's space launch capabilities.

It was a successful launch, which an elated Vajpayee viewed from a specially constructed enclosure. Within the sprawling launch complex at Sriharikota on the Bay of Bengal coast, viewers are generally accommodated on the terrace of the Brahm Prakash Hall (named after Prof. Brahm Prakash who was a director of the Vikram Sarabhai Space Centre in Thiruvananthapuram). This hall is located nearly five kilometres from the actual point where the rocket lifts off. It is a sight that defies description – a giant rocket streaming above lanky eucalyptus trees that otherwise block one's view, the roar of its powerful engines and the incredible smoke trail. It is an image that remains with viewers long after the rocket disappears into the sky.

In the ISRO tradition, a post-launch press briefing was held in the Brahm Prakash Hall, which was packed to capacity with journalists from all over India. A visibly thrilled team of scientists and engineers accompanied Vajpayee. The first question was asked by Pallava Bagla, co-author of this book: would the government fund a mission to the moon? While there was surprise all around that the first question had little to do with the launch just witnessed, a somewhat taken-aback Vajpayee turned to ISRO officials on the dais for a quick consultation. Meanwhile, Kasturirangan jovially remarked that the PM would surely have something to say about the moon, poet that he was. Vajpayee, in finally answering the question, did take refuge in his reputation as a poet: he said in his usual and somewhat quirky style that those who have landed on the moon have found it does not have a beautiful face. A successful launch, laughter all around, and a neat ducking of an important question – there was something there, and something vital. That much was clear.

The copy-book launch off Sriharikota and the other launches before and after testified that India now had launch and satellite technology under its belt. This is why on 11 May 1999, when speaking on 'India's Space Odyssey', Kasturirangan had said that India could easily launch a small satellite of about 275 kg in a flyby mission to the moon, or even place a 140-kg satellite in orbit around the moon. The objective would be to study the moon's core.

The milestones in 1999, when Indians first came to know of India's grand plan, were significant for two important reasons. First, a mastering of cutting-edge technology – there is hardly any argument on the fact that technology defines development, and space can become the next frontier for human development. Second, if these developments could excite young people and attract them towards the study of space science and planetary exploration, the country would see an infusion of fresh talent into one of the most critical areas of scientific research and development. So that in a classroom at any average primary school, when the teacher asks the routine question – 'What do you want to become when you grow up?' – there will be more than the one kid who says 'astronaut' without having everyone in the room stare and giggle.

One could say that the idea of a mission to the moon was born in the early 1960s, with Thumba. That was the time when space research activities began to be formally planned, with the initiation of programmes to study 'ionospheric and electrojet phenomena'.[5] The early focus on creating technologies capable of fulfilling social needs then began to change. When ISRO started to feel it was necessary to shift to research-oriented outer space missions, a strong body of criticism emerged. Dr George Joseph, director of the UN-affiliated Centre for Space Science and Technology Education in Asia and the Pacific (CSSTEAP), Dehra Dun, and honorary distinguished professor, ISRO, interprets this shift by explaining how 'it was time to start defining a second vision for the Indian space programme'. In the late 1990s and early 2000s, it was evident that after three decades spent focusing on the earthly pursuits of weather forecasting and communications, India's space agency was raising the bar.[6] These were the years when the bricks and mortar were laid for the foundation of India's lunar mission.

At the sixty-fifth annual meeting of the Indian Academy of Sciences in October 1999 in Lucknow, a symposium was held, titled 'An Indian Case for Going to the Moon'. Kasturirangan and other leaders of the space science community had been speculating since 1997–98 that if Indian remote sensing satellites were generating such top quality images, perhaps it was time to open a new chapter in scientific and exploratory missions. Lucknow somehow seemed the perfect fit as a venue – the city's role in sparking the mutiny of 1857, which led to the first war of independence, is legendary. And this meeting, in more ways than one, was the initiation of another revolution – a technological one.

In the otherwise impressive premises of the Central Drug Research Institute at the Chatter Manzil palace, the narrow hall where the meeting was convened left much to be desired – it was hard to darken in daylight and equipped with an old-fashioned overhead projector which was what was used to announce to the Indian scientific community that ISRO was willing to undertake a mission to the moon, provided the Indian scientific community thought it worthwhile. Most of the information shared and discussed by the 200-odd space scientists and researchers gathered together pointed at the fact that India was not only willing, but also technologically ready to set out on the journey. Interestingly, the meeting didn't quite elicit the kind of excitement that one would have expected, perhaps because Indian scientists had not yet given serious thought to the possibility of such a mission. But there was no overt opposition either.

To take the plan to its next level, Kasturirangan requested Dr Narendra Bhandari to step in and chalk out the scientific goals for a moon mission. The following year, in February 2000, the Astronautical Society of India (ASI) met at the Space Applications Centre (SAC) in Ahmedabad, and drew up formal recommendations for the scientific exploration of space.

Soon after, a National Lunar Mission Task Force was constituted by ISRO with Joseph, who was then director of SAC, in charge. Leading from the front, he steered to closure a report that formed the basis for a final decision and approval by the Indian government. 'This was a national effort, and a national team,' says Joseph, who staunchly believes it is Kasturirangan's unwavering commitment to an inclusive and national approach, including his decision to guage the mood of scientists across the country, that made the exercise a success. Joseph is also quick to credit the 'very dynamic' member secretary of the Task Force, Dr K. Thyagarajan, who made significant contributions to the activities of the group.

Coming up with an 'assessment on the feasibility of an Indian mission to the moon as well as dwelling on the focus of such a mission and its possible configuration', the Task Force also 'provided specific inputs such as the primary scientific objectives of such a mission, plausible instruments to meet these objectives, launch and spacecraft technologies that need to be developed and the need for setting up of a Deep Space Network (DSN) station in India for communication with the lunar orbiting spacecraft. The team also provided a provisional budgetary estimate.'[7] Joseph says, 'I knew two things right from the start, and with great clarity. This would not be a "*humne bhi kiya*" mission at all, it would be innovative. Second, I thought it was totally doable.'

In 2003, a hundred well-known scientists from fields as diverse as earth, planetary and space sciences, astronomy and astrophysics, engineering and communication sciences, physics and chemistry got together in Bangalore to review the study report of the Task Force. Reflecting confidence in ISRO's technical capacity to execute the task, there was a unanimous recommendation that India should undertake the mission. The resurgence of global interest – with several nations planning similar journeys – was a critical factor that contributed to the group's final recommendations. There was also an understanding that the Indian mission would be a shot in the arm for basic science and engineering research in the country, while setting up new challenges for ISRO. Besides, it could become a magnet for young talent – the lack of which had been felt in several areas of basic sciences and fundamental research for some time.

This was an exciting time: decision-making was obviously on the fast track, spurred as it was by the knowledge that worldwide, countries, governments and scientific communities would sit up and take notice of India's lunar mission plan. The reputation and credibility of ISRO – known for delivering on its promises – also helped speed up the decision. In May 2003, the Parliamentary Review Committee gave its go-ahead, and papers were put together to seek central clearance. The mission plan received the government's stamp of approval through the Cabinet in November 2003. Thus, the 24-month mission, expected to create the first digital elevation map with a high spatial and altitude resolution of five metres, was put on track. ISRO explained that it was to be a purely exploratory scientific mission and the participating institutions would include the Physical Research Laboratory, Ahmedabad, the National Physical Laboratory, New Delhi, the Tata Institute of Fundamental Research, Mumbai and the Indian Institute for Astrophysics, Bangalore.

While most scientists were celebrating the quick approval of the mission by the government, some openly expressed their concern, describing it as a profligate step by a poor nation. Some critics felt that the shortage of young researchers keen to take up a career in astronomy or astrophysics was a hurdle. Senior researchers like Dr Jayant Vishnu Narlikar, a theoretical astrophysicist and former director, Inter University Centre for Astronomy and Astrophysics, Pune, emphasized that a lack of trained personnel could hamstring the programme if remedial measures were not taken urgently. Narlikar stressed, 'You need young and dynamic researchers to sustain a programme as farsighted as this, and for that a dialogue has to be initiated immediately and in right earnest with the universities.' Dr H.S. Mukunda, a senior aerospace engineer at the Indian Institute of Science, Bangalore, also voiced the opinion that trying to do something other countries had done a long time ago would bring no technical benefit.[8]

Dismissing such criticism, Kasturirangan said, 'Chandrayaan-1 is expected to be the forerunner of more ambitious planetary missions in the years to come including landing robots on the moon and visits by Indian spacecraft to other planets in the solar system.'

This view was echoed internationally. As James C. Dodge, India specialist working in the Office of Earth Science, NASA Headquarters, Washington, DC remarked, 'The money could alternatively be spent on aid to the needy, but a certain measured amount of the Gross National Product should be used to thrill the imagination of everyone. I think the whole global community would like to see the emergence of another moon-faring nation. It certainly would demonstrate unequivocally that India has the technological capability to undertake very challenging missions.'

As Jayati Datta and S.C. Chakravarty of ISRO put it:

> Chandrayaan-1, the first Indian mission to the Moon, would play a major role in reinvigorating research in fundamental science, help in upgrading technological capabilities for future space systems, become a meaningful contender in the international space arena for exploration and commercial exploitation of resources in the immediate neighbourhood of our planet. Above all, the mission would help young talented scientists and engineers of the country to take up the challenging high level of intellectual activities in basic sciences. Chandrayaan-1 mission would pass on the baton to sophisticated future lunar and planetary missions with possible landing and sample return capability. [9]

The overarching mandate of the mission is quite unambiguous – to expand the country's scientific knowledge about the moon, and contribute to the world's pool of knowledge that can then translate into advancement for humankind. Riding on this is a clear and strategic objective of being able to upgrade the country's technological aptitude and wherewithal – significant in the current global context, when earthly matters will start to depend more and more on the rest of space. As Madhavan Nair says:

> As far as humankind is concerned, yes, we have a lot of problems to solve on the planet earth. But at the same time the next frontier for humankind is going to be

outer space. For various purposes, whether scientific observations, manufacturing, power generation or even from the security angle, outer space is going to be the most important location. So our presence in outer space will become a logical extension of any space research activity we undertake.

The Science and the Technology

Chandrayaan-1 is a robotic mooncraft that travels using a modified version of the PSLV. It is expected to generate high resolution, three-dimensional maps of not just topography but its chemical composition and the distribution of minerals. In fact, Chandrayaan-1 is a remote sensing satellite on a resource-mapping project. By reading the maps, scientists hope to understand better the complex nature of the origin and evolution of not just the planetary system but also the moon-earth system. Using visible, near infrared, low energy X-rays and high energy X-ray regions, the probe will conduct a complete remote sensing exercise on the moon's surface.

The project director for the satellite of the lunar mission is the dynamic Mylaswami Annadurai; he oversees a team of over 200 Indian scientists and engineers. Annadurai, an electronics engineer, was just 46 when he was tasked with this significant mission. Officials at ISRO said then that the choice of a young team leader was to ensure continuity in India's interplanetary exploratory ventures. Born to an elementary school teacher, Annadurai has often said that Chandrayaan-1 is the most exciting mission that he has worked upon 'because new frontiers have to be broken'. As a young boy growing up in Kothawady, a village off Coimbatore, Annadurai had thought

Mylaswamy Annadurai, Udaipur, 2004

that the sky above his village was like a natural planetarium; today, he is completely absorbed in one of India's most prestigious space exploration projects.

It was at the sixth international conference on the exploration and utilization of the moon in Udaipur, Rajasthan, in November 2004, that the first scale model of the cuboid Chandrayaan-1 was displayed for people outside ISRO to see. After a five-and-a-half-day journey of about 4 lakh km from the earth to reach the moon, Chandrayaan-1 is to stay in a 100-km polar orbit around the moon.

It was Prof. Bhandari who crafted the scientific objectives of the mission. For him the power of scientific knowledge is a moving force, and it was also the key motivation to mount a mission to the moon. Now, Dr Jitendra Nath Goswami, Director, PRL, is in charge of all the scientific activities of the moon mission.

The Indian lunar probe is expected to provide a three-dimensional atlas (of high quality, given that the images would have a 5-10 metre spatial and altitudinal resolution) of the near and far sides of the moon. Such a high-resolution map would be useful since the moon would have important historical details of planetary evolution hidden in its unexplored surface and core. As we know it, the moon has no atmosphere, which is why there is very little wind erosion. Features as old as the moon itself are likely to be preserved in their original form, and analyzing these features may help us understand better how the earth came to be what it is today, or more likely what the earth was like before life evolved on it.

Generating maps of the geological units of the moon and mapping the lunar surface based on its chemical composition – trying to locate and map magnesium, aluminum, or other chemical elements like silicon, calcium, iron and titanium – will also help piece the jigsaw together. Essentially, understanding the geological characteristics of the moon would mean a clearer picture of planetary formation, and the evolution of the solar system. We would be able to learn about the processes that led to the moon's formation and the different chemical processes and reactions that took place during the early days of its formation. Space scientists and technologists, even entrepreneurs, are hankering after knowledge about the resources available on the moon, such as Helium-3, which could be a potential energy source if and when fusion energy becomes a reality at the International Thermonuclear Experimental Reactor (ITER) being built in France, a project in which India is an equal partner. The resources on the moon are clearly an important reason for the global scramble.

According to ISRO scientists, 'The chemical maps will enable [us] to determine the heterogeneous nature and stratigraphy of the moon's crust and thereby test certain aspects of the magma ocean hypothesis and may allow [us] to determine the compositions of impactors, which are also relevant to the formation of the Earth.'[10]

One of the theories behind the formation of the moon is that a large amount of energy was liberated when it was formed (the Giant Impact Hypothesis) and that because of all that energy, large portions of the moon remain molten – therefore the term 'magma ocean'. Studying mineral compositions on the lunar surface may give scientists an idea of the timelines, as to when processes like melting or crystallization occurred.

For many years, in the pre-space exploration era, the only lunar 'science' was focused on orbits, density, momentum and other related aspects. And of course there were theories, largely tentative and speculative, of the origins of the moon – such as the Giant Impact Hypothesis. There is hope that the missions being planned now will fill in the blanks, and enhance knowledge about the moon. 'The objective of the lunar mission is to derive maximum scientific knowledge about the moon' is a statement often repeated by Madhavan Nair.

Prior to the Udaipur conference in 2004, the plan was only to orbit the moon. At this meeting, something transpired between ISRO and Dr Kalam, himself an aeronautical engineer of high repute and the man credited with the mastery of the forerunner of PSLV. In Udaipur, Kalam voiced the idea that Chandrayaan-1 should not merely be an orbiting mission but that there should be a lander as well. Easier said than done, but a suggestion readily conceded to by Madhavan Nair, which is how the Moon Impactor Probe came into being as a

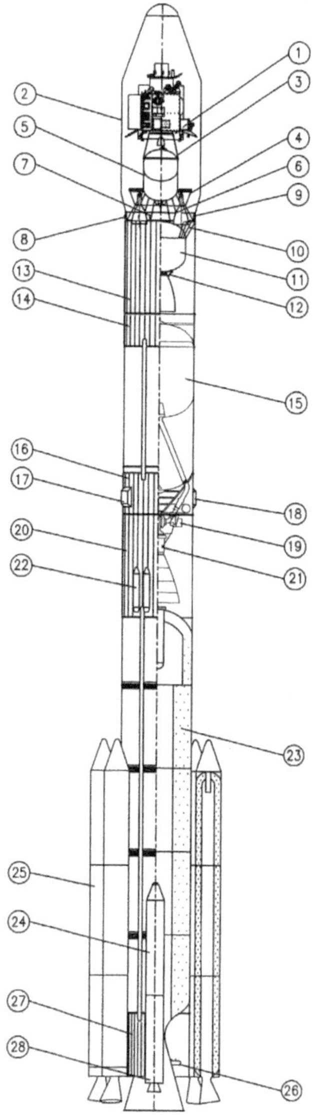

PSLV : The Moon Rocket

proposal to supplement the basic mission idea. A soft lander was difficult to achieve in this short duration, so an impactor was decided upon. This meant that while it would still be an orbiting mission, India would be sending down a probe to the lunar surface, which would assess the lunar atmosphere and crashland on the moon.

Within a month of reaching the designated lunar orbit, the Moon Impactor Probe, piggybacking on Chandrayaan-1, would detach from it and head moonward to drop the Indian tricolour on the moon's surface – doubtless an event of great geo-political significance. A 29-kg instrument the size of a computer monitor, fabricated by scientists at the VSSC, the impactor, after leaving the mother ship, would rapidly descend towards the lunar surface. In its 20-minute flight, it would capture high-resolution photographs of the moon's surface, analyze the sparse lunar atmosphere and transmit back to the orbiting Chandrayaan all the data for further transmission to earth. The impact probe would have three payloads – a radar altimeter, a video imaging system and a mass spectrometer. With this Rs 4 crore gadget, ISRO hopes to master technologies needed for landing on the moon while also looking for a suitable landing site, preferably closer to the lunar pole where scientists suspect the presence of ice.

India already has a proven capability to build sophisticated remote sensing cameras.[11] Five Indian instruments including a Terrain Mapping Camera (TMC) that can capture stereo images, X-ray mapping cameras and spectrometers are part of Chandrayaan-1. The TMC will map the complete topography of the near and far sides of the moon, and the three-dimensional atlas will be a result of this image capture.

Meanwhile, another payload, a Hyper Spectral Imager (HySI) will focus on generating a mineralogical map of the moon using the scientific technique of spectral separation. There is also a High Energy X-ray spectrometer (HEX) designed to study emissions on the moon due to the radio decay of uranium and thorium. A Lunar Laser Ranging Instrument (LLRI) will use pulses of light to illuminate the surface of the moon and study scattering, absorption and re-emission through a detector, thus providing details of the moon's topography.[12] All these instruments have been accommodated on what is called a one-ton satellite bus that ISRO has used for building Chandrayaan-1. The many complex units of this bus must work in unison for the two-year mission to be successful in its entirety. Not an easy task, for several complicated systems have to function in a sequential and parallel manner over a very long period of time in the hostile environment near the moon.

A significant spinoff of the Chandrayaan mission is the infusion of fresh interest among the young. Today, Bhandari tells many young students and researchers: 'If you must study chemistry, think of lunar chemistry, if you want to study geology, why not lunar geology?' The pursuit of scientific knowledge is also about defining a challenge and seeking solutions to it – baffling though it can be for communities for whom even the search for basic amenities is a challenge. But not too distant in time is the possibility (no longer hazy) of a global community reaching into outer space for diverse developmental needs. Against that backdrop, it becomes so much easier to decode the logic and motivation behind India's plans and journeys into outer space. As Narlikar wrote in the *Times of India* in 2003, 'There is also the intellectual challenge that comes now and then and prompts an individual or a

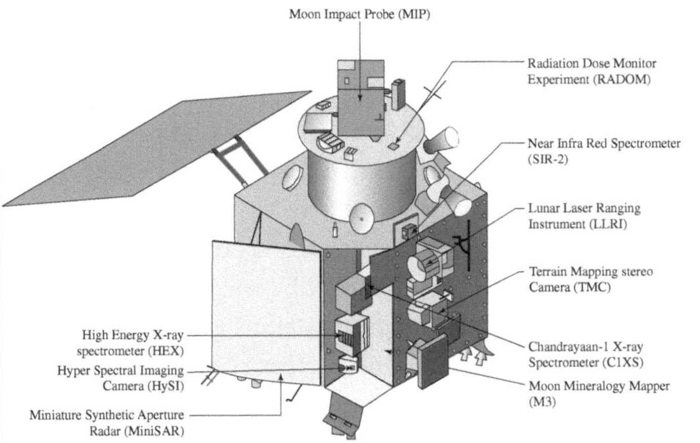

The Chandrayaan-1 satellite
Courtesy: ISRO

civilization to rise and meet it. India has a tradition of taking up such challenges, vide its stake in Antarctica. You climb the Everest because it is there. You go to the moon because it is there.'[13]

Clearly, a scientific mission such as this can become a reality only if it is backed by sophisticated technology. India's maiden unmanned mission to the moon is to be launched atop ISRO's locally made workhorse, the 44.4-metre, 316-ton PSLV, in its improved and more powerful avatar, PSLV-XL. Designed and developed at VSSC with various developmental aspects taken care of at the ISRO Inertial Systems Unit (IISU) and the Liquid Propulsion Systems Centre (LPSC), also in Thiruvananthapuram, the PSLV had its first successful flight in October 1994. It is a launcher with a great record – 12 consecutively successful missions and just one failure – and it has consistently demonstrated multiple satellite launch capability. International customers have used the PSLV, and 16 satellites have been sent up into space for overseas customers. Other than that, the PSLV has launched 13 Indian satellites – nine of the IRS satellite family and one, a weather watcher, or a meteorological satellite, besides HAMSAT, a small amateur radio communications satellite, a 550-kg recoverable capsule called Satellite Recovery Experiment-1 (SRE-1) and the Indian mini satellite in 2008. PSLV was also used to launch ISRO's exclusive meteorological satellite, KALPANA-1, into a geosynchronous transfer orbit in September 2002. The vehicle in its PSLV-XL form will be used to launch Chandrayaan-1.

The basic idea was to have a launcher for ISRO to place 1,000-kg IRS satellites into 900-km polar sun-synchronous orbits. This basic capability of PSLV has since been enhanced to 1,600 kg. According to ISRO, various innovations and improvements such as increased propellant loading in the stage motors, using composite material for the satellite mounting structure and changing the sequence in which the strap-on motors get fired up, have helped make PSLV a better rocket.

Every rocket has stages and the PSLV has four, which use solid and liquid propulsion systems alternately. The first is among the largest solid propellant boosters in the world and carries 138 tons of propellant. Strapped to the core stage are six solid propellant strap-on motors with nine tons of solid propellant each. The second stage has 41.5 tons of liquid propellant. The third stage uses 7.6 tons of solid propellant and the fourth has a twin-engine configuration with 2 tons of liquid propellant. There is also a 3.2-metre-diameter metallic bulbous payload fairing that is meant to protect the satellite when it's on board the rocket; this is discarded once the rocket clears the dense atmosphere. Stage separation, payload fairing separation and other needs are fulfilled by auxiliary systems. There are sophisticated systems to control the vehicle and guide it through the pre-determined trajectory. The vehicle performance is monitored through telemetry and tracking.

The venue of the lift-off is the Satish Dhawan Space Centre that houses the twin launch pads used by India to send rockets into space at Sriharikota. The entire verdant island, separated from the mainland only by a very thin strip of water, is occupied by ISRO and houses some of the most sophisticated facilities in India.

The spacecraft was built at ISRO's Satellite Centre in Bangalore. The satellite weighs about 590 kg at lunar orbit and has a solar panel array for power. It also has space for a scientific payload of over 100 kg. A bipropellant integrated propulsion system is being used to carry the satellite into lunar orbit. Using three Solid State Recorders (SSRs), the space craft will be recording all the data that flows in from the various payloads; a parabolic dish X-band antenna 0.7 metres in diameter will allow for the transmission of data from payloads.

All this action more than 400,000 km away from earth requires a massive network of technology and equipment on the ground. A sustained and continuous radio link is one of the most critical aspects of any mission such as Chandrayaan-1, and the operational control of the mooncraft is anchored within the Indian Deep Space Network (IDSN), a ground station and the network that supports Chandrayaan-1. There is also a Spacecraft Control Centre (SCC) and the Indian Space Science Data Centre (ISSDC). A huge 32-metre antenna at Byalalu, a little known village about 40 km outside Bangalore came up late in 2007, joining another antenna of 18 metres in the same region. These two are important elements of the IDSN, not just for Chandrayaan-1, but also for future missions and initiatives.

Byalalu, nestling as it does in a basin-like region within the Deccan plateau, is a strategic choice for the control centre, with minimal electromagnetic interference. For ISRO's Telemetry, Tracking and Command Network (ISTRAC), the IDSN will be a hub for the moon mission, through which mission controllers will communicate with the spacecraft. Ensuring that all is well with the spacecraft, maintaining its orbit and altitude to the requirements of the mission, conducting payload operations and making data available to all concerned will be the responsibility of the ground segment of Chandrayaan-1. In fact, a major chunk of the total estimated budget of Rs 386 crore for Chandrayaan-1 is for the IDSN, pitched at Rs 100 crore.

Funds have not been a constraint for ISRO and its several programmes because the Indian government, for the most part, has not been tightfisted. The PSLV budget is estimated at another Rs 100 crore. The cost of developing the payloads is roughly Rs 53 crore and the spacecraft bus comes at a cost of Rs 83 crore.[14] ISRO was alloted a fifth of its anticipated cost of undertaking the Chandrayaan-1 mission in 2004.

The Partners

Back in the 1960s, Sarabhai had said that India was far from living any 'fantasy of competing with the economically advanced nations in the exploration of the moon or the planets or manned space flight'. With growing technological sophistication, India today seems to have come full circle: in

2004, it announced its intention to make the moon mission an international affair, the basic idea being that other countries could get a free ride in return for sharing scientific data. This was an extraordinary and unprecedented gesture, reflecting the maturity of the Indian space community. Nair is known to have said then that 'we will not charge for flying the payload from others on board our spacecraft for Chandrayaan'. This translates as India's willingness to foot a bill of over Rs 51 crore (the estimated cost for a 10-kg scientific instrument) on behalf of international partners. Nair had then said that international proposals would be evaluated on the basis of their scientific objectives and the degree to which they could complement and supplement the Indian instruments.

In February that year, ISRO announced that there was room for a 10-kg payload on its orbiting lunar probe and that Canada, Russia and Israel had already expressed interest[15] in becoming active partners on the Indian mission. At the time, Washington was still strident about the sanctions it had imposed on ISRO after Pokharan. But NASA's Dodge was quoted as saying that 'ISRO has the technical capability and the management excellence that is required to mount and succeed in a moon mission.'

In June 2004, the India–United States Conference on Space Science, Applications and Commerce was organized by the ASI and the American Institute of Aeronautics & Astronautics (AIAA). This was the first time Indian and American space scientists were meeting face to face after 1998. The focus was on strengthening cooperation in space science and its applications.

It wasn't an easy time for India, especially with regard to relations with the US. Post-Pokharan, sanctions had been

Dr Spudis (left) and Dr Annadurai (right) with other Indian and American officials, discussing details of the Indian moon mission, Udaipur, 2004

imposed on several Indian institutions conducting fundamental and strategic nuclear research. ISRO was struggling to import simple computer chips, the kind now used in everyday laser printers. The American policy of denial of all exports to ISRO hampered rapid progress. But there was a bright side – denied fish, so to speak, India decided to step out and learn how to fish. Of course there was a cost to it – heavy time and cost overruns on vital projects like the development of a cryogenic engine, a project that suffered a decade's setback.

At this time of extreme technological hostility towards India from the developed nations, it was a significant move for ISRO to solicit proposals from the international space community. Later, in November 2004, over 200 representatives from across the world gathered in Udaipur to figure out the future contours of lunar exploration. The meeting was used as an occasion by ISRO to bridge gaps with the international teams that were offering payloads on Chandrayaan-1. There were also several meetings on the sidelines with the US delegation.

The main meeting took place in the majestic Durbar Hall of the Fateh Prakash Palace, adorned with large, imposing portraits of the Mewar maharanas. Wherever you looked, there was grandeur as typified by Rajasthani royal culture: weaponry and paintings, chandeliers and velvet canopies, all with gold inlay and glitter. On the first floor was the Crystal Gallery, a vast collection of precious crystal said to be among the largest private collections in the world.

When President Kalam arrived for the inauguration of the meeting, he was seated on a silver throne, while the others were on plush chairs. There was confusion on the stage when he got up and asked for the throne to be removed. He wanted to be given a chair like everyone else. 'I am within the scientific community which is my fraternity,' he said.

At the Udaipur conference, there was overwhelming international participation – in some ways an indication of the level of interest worldwide in India's plans and scientific aspirations. Seventeen countries were represented and this made the Udaipur platform a robust international collaboration and partnerships meeting.[16]

Excerpts from the Udaipur Declaration, November 2004

A new lunar decade has begun. SMART 1 technology opens the fleet of new missions being flown to the moon. Plans for Chandrayaan-1 are timely, specifically relevant, building infrastructure for future exploration. Additional complementary missions of exceptional value include Selene, Lunar A, Chang'e and Lunar Reconnaissance Orbiter.

We acknowledge that fundamental science questions about the moon remain to be addressed, not only to understand the early history of the Earth/moon system and its current environment, but also to acquire knowledge for the next steps of exploration and human utilization. Of prime importance is the formation and evolution of the terrestrial planets, including the origin of the moon. Also vital is the impact history at 1 AU (astronomical unit), including the absolute timing of early events such as the giant basins. A major unknown is the internal structure of the moon, both its geophysical and compositional properties. The moon is a natural laboratory for studying its interactions with the space environment, together with the products resulting in the polar deposits.

Recognising that any lunar exploration program must later include advanced orbital instruments as well as in situ analyses from several surface stations and targeted sample return, we urge a broad and open discussion and coordination for the selection of landing sites to optimize the science return and benefit for exploration.

We believe that exploration and utilization of the moon will bring benefits to human kind as well as serve national needs, and we recommend an international plan for its implementation. The participants endorsed the ILEWG stepwise approach, starting with a joint science analysis from ongoing precursor missions (SMART 1, Lunar A, Selene, Chang'e, Chandrayaan -1, Lunar Reconnaissance Orbiter) continuing with lunar landers, and cooperating into an international lunar robotic village before 2014, evolving technologies for man-tended missions and preparing the ground for an effective, affordable human lunar exploration and permanent presence by 2024. We encourage space agencies to coordinate and integrate their plans in a robust international moon-Mars roadmap in coordination with the ILEWG roadmap, where partners can identify their contribution for an effective implementation using their skills.

As we move forward with mission implementation, we urge space agencies to study and coordinate international lunar infrastructures and assets, such as telecommunication, navigation, logistics, lunar internet, towards effective lunar exploration. We specifically recommend coordination of international efforts for the establishment of standards to facilitate lunar exploitation and settlement – e.g. use of the metric system, well-characterised lunar soil simulants, common data formats and instrument interfaces; frequency, and power. We also urge the establishment of a standard lunar geodetic network, and recommend that the moon Treaty be revisited, refined and revised as necessary in the light of the present-day impetus for expeditions, both robotic and human, to the moon by several nations.

> We believe missions to the moon have an enormous potential to inspire both the young as well as their parents because the moon is visible to all and is within our reach to visit. Herewith we encourage young scientists of different fields and nations to join this activity and work together in realizing lunar exploration goals.[17]

The ensuing months were hectic and filled with interactions and formal discussions with global partners. In May 2006, Dr Michael Griffin, NASA's chief administrator, formally inked an agreement with ISRO that would put two sophisticated US instruments on India's mission to the moon. What may have been a small step for the Americans became a giant leap for the Indians. It also opened a new chapter in Indo–US space relations that had hit rock bottom by then.

The well-known American inter-planetary scholar Dr Paul Spudis, at the Applied Physics Laboratory at Johns Hopkins University, is the principal force behind one of the payloads on Chandrayaan-1 called MiniSAR (Miniature Synthetic Aperture Imaging Radar Instrument), aimed at detecting and confirming the presence of water ice in the lunar polar regions up to a depth of a few metres. It will map polar ice deposits on the moon, first hints of which were thrown up by the lunar mission Clementine in 1994. Though, scientists still don't agree about the presence of water ice – those who do, disagree on its volume, thickness and composition. According to Spudis, 'the imaging radar can "see" into the dark regions of the poles and map the ice by its distinctive radio properties.'

The other American instrument is a Moon Mineral Mapper (M^3) designed by Dr Carle Pieters of Brown University

American lunar specialist Dr Paul Spudis, Udaipur, 2004

and meant to trace, among other things, the richest sources of Heluim-3.

The payload from Bulgaria is called RADOM (Radiation Dose Monitor Experiment), and it was proposed by Dr Tsvetan Dachev at the Solar-Terrestrial Influences Laboratory at the Bulgarian Academy of Sciences. It seeks to measure solar wind particle flux, the energy spectrum and radiation dose map around the moon.

Dr Stas Barabash of the Swedish Institute of Space Physics, Sweden, in collaboration with Dr Anil Bhardwaj of VSSC, are to fly SARA or the sub-KeV atom reflecting analyzer, with the primary objective of imaging magnetic anomalies and lunar surface composition in order to locate volatile rich areas. Barabash says the payload will conduct 'a novel experiment never performed before'.

Dr Urs Mall, from the Max Planck Institute for Aeronomie, Germany, sent an instrument called SIR-2 (Near Infrared Spectrometer) whose main objective is to study in detail the geological and mineralogical aspects of the lunar surface from a circular polar orbit. Mall says the experiments may throw light on 'the cause of the global asymmetry of the moon as part of which the far side has (a) thicker crust as compared to the near side.'

British scientists from the Space Physics Division of the Rutherford Appleton Laboratory, UK, have sponsored an instrument called CIXS (Chandrayaan-1 X-ray spectrometer), a low energy measuring device whose primary goal is to carry out high quality X-ray spectroscopic mapping of the moon using X-ray fluorescence spectra. Dr Manuel Grande, who initiated this proposal, said he hoped that 'the measurements would elucidate the key questions on origin and evolution of the moon'.

For many of these experts, collaboration with India is reflective of growing opportunities for flying 'Western' technology on 'Eastern' space missions. For the teams in India, given the amount of hard work that has gone into the mission, Western interest is not surprising. In an article in the *Journal of the Aerospace Sciences and Technologies* in 2006, Annadurai stated: 'The Indian mission to the moon should be seen beyond the scientific results it produces. Studies have shown that the moon could serve as a source of economic benefit to mankind and could be of strategic importance. The moon can be both, a beacon and a focus for the next generation of space exploration which will accrue new and important benefits, to the people of all nations and the earth. Just a few decades back, man never imagined that he would set foot on the moon. Decades from now, human colonies on the moon could become a reality.

India should also be in the forefront of this challenging and exciting endeavour. Chandrayaan-1 is the first calculated and well planned initiative by ISRO in this direction.'[18]

There is no doubt that the mission also allows for international parity – of no little significance in these fraught times. There is also little doubt that Chandrayaan-1 presents unique prospects of cutting-edge scientific planetary research. India sees, as many other nations do, the potential in staying on top of this particular dimension of global development – who knows which spatial way the future will beckon?

Besides the scientific value of such collaborations, the idea of flying Western technology on Eastern space missions is of strategic value. As a country and as a people, India and Indians have always been excited by the idea of self-sufficiency, and Chandrayaan-1 promises exactly that. As Joseph says, 'In future when development takes place in space, I hope Chandrayaan-1 maps will be used.'

6

The Future Is Here

The most incomprehensible thing about the universe is that it is comprehensible.

– Albert Einstein

First machines, and then men and women too, who knows? Not only India, many nations around the world are moonstruck.

When the annual budget of the Indian government for 2007–08 was announced, Union Minister for Finance P. Chidambaram highlighted the allocation of Rs 50 crore for ISRO's manned space initiatives – a possible manned spaceflight in 2015 and a human on the moon by 2025.[1] The manned mission is expected to be a 'fully autonomous orbital vehicle carrying 2–3 crew members to a 400 km low earth orbit, safely returning them to earth, with ISRO playing the lead role.'[2]

In the next eight to ten years, this plan will roll out, based on exhaustive feasibility studies. According to Dr B.N. Suresh, former director of the VSSC, the studies have been promising.

Dr Madhavan Nair and Mr A. Perminov, Director, Roskosmos, sign an agreement in Moscow, 12 November 2007

The GSLV Mark II, an existing vehicle, can be used to carry humans into orbit and bring them back –which is not to say that other required technology is ready. 'Hopefully we feel we will be able to do it by 2015,' says Suresh.

The entirely indigenous project is expected to cost close to Rs 10,000 crore. Standing tall at 50 metres, and weighing 400 ton, the GSLV Mark II is currently being fine-tuned to take on the task. Meanwhile, through the successful SRE project in 2007, Indian scientists managed to test in real-time the very critical thermal protection mechanism that allows a safe return, with Indian-made heat resistant tiles. Prototype drawings of the crew module are also in progress. India now seems set on entering the league of nations that have manned space flight capability.[3]

In November 2007, India and Russia signed an agreement to explore the moon together. They plan to develop a robotic rover that will be deployed from a lunar orbiter to probe the moon's surface for geological data and look for Helium-3.[4] ISRO's Nair, and A. Perminov, Director, Roskosmos (Russia's Federal Space Agency) signed the agreement envisaging cooperation to send into space Chandrayaan-2, a joint lunar mission.

> ISRO will have the prime responsibility for the Orbiter and Roskosmos will be responsible for the Lander/Rover. A few scientific instruments from other space agencies may also be accommodated on these systems. Chandrayaan-2 will be launched on India's Geosynchronous Satellite Launch Vehicle (GSLV) around 2011–12. This agreement is a major milestone in the long-standing cooperation between India and Russia in the area of outer space.[5]

Prime Minister Manmohan Singh called it an 'area of great promise' for the two countries. The rover will conduct in situ analysis of moon rock and soil, and transmit data to the spacecraft, which in turn will beam data to the ground network antennae.[6] The two countries have been traditional partners in this business of exploring outer space – it is in a Russian rocket, Soyuz, that India's only astronaut to date, Rakesh Sharma, travelled to space in 1984. India's heavy communications satellites have also been transported into space on the GSLV using Russian-made cryogenic engines.

The future certainly seems action-packed: there are at least two dozen missions planned before 2017. While the first phase of the lunar race had just two participants, the US and

Russia, several countries are now heading into deep space – not just to the moon but beyond, into the rest of the solar system. Starting with robotic missions, first with satellites, then rovers and landers, followed by sample returns, these projects will finally see humans landing on the moon and beyond. Most countries are talking of international collaboration, but firm plans beyond the fifteen-nation International Space Station are still to be finalized.

Russia has a three-step space programme for the future, according to Pernimov. The first step for Roskosmos is to integrate a segment into the ISS and set up a multipurpose laboratory by 2015. Between 2016 and 2025, Russia plans to continue to integrate its complex into the ISS and build up a space crew. From 2026 to 2040, the idea is to develop near-earth space infrastructure, which would form the basis for human flights to the moon and Mars.

Countries like India, China, and nations within the ESA are hoping to receive approvals for placing humans on the moon. Meanwhile, for a country like Japan, the focus is on the continued exploration of the moon, since the space community, led by Dr Keiji Tachikawa, president of the Japanese Space Exploration Agency, feels that the moon should no longer be a place for a mere visit; it should be seriously considered for human habitation and for the potential of its mineral resources.

Kasturirangan has often alluded to the fact that Chandrayaan-1 is only the beginning of a serious and active lunar phase for India's space programme: 'Chandrayaan-1 is expected to be the forerunner of more ambitious planetary missions in the years to come including landing robots on the moon and visits by Indian space crafts to other planets in the solar system.' Yet, the core interest for India still remains social application. At the close of 2007, the Indian magazine *Current Science* published a special issue on the Indian space programme, and Kasturirangan wrote convincingly in the editorial: 'In the context of the strategy for the management of the future of India's space programme, in our considered view the social objectives will continue to be the main driver.'

Outpost of the future

'As civilization spreads throughout the solar system, the moon will provide the main link between the Earth and her scattered children,' Dr Kalam said at the Udaipur conference in 2004. He explained how the moon, given all its characteristics, would become a telecommunications hub for interplanetary communications. With low gravity, a perfect vacuum and the frequency-less moon sky, besides temperature extremes, it would present a perfect environment for waves of all frequencies.

The International Lunar Exploration Working Group (ILEWG), a public forum created in 1994 and sponsored by the world's space agencies, is training the spotlight through intense brainstorming, on 'international cooperation towards a world strategy for the exploration and utilization of the moon'. At the group's ninth global conference, participants re-dedicated themselves to the cause: 'We reaffirm our commitment to international lunar exploration, from the science analysis of current precursor robotic missions, to the global robotic village, and the preparation of international human settlements on the moon.'[7] ILEWG meets regularly, at least once a year, and leads the organization of an international conference to discuss the state of lunar exploration. Formal reports of its work are presented to the Committee on Space Research (COSPAR) and to space agencies.

The aim of the ILEWG is:

- To develop an international strategy for the exploration of the moon
- To establish a forum and mechanisms for the communication and coordination of activities
- To implement international coordination and cooperation
- To establish an electronic communications network for the exchange of science, technology and programmatic information related to lunar activities, in order to facilitate communication among all interested parties

Whatever the means of getting there, the future seems to hold images of a base camp on the moon, meant for refuelling as missions move further away towards Mars. Dr Michael Griffin of NASA is convinced that the first human being will land on Mars in 2037. Speaking at the International Astronautical Congress in Hyderabad in September 2007, he said, 'When this conference meets in 2057, I am convinced that we will be able to celebrate the 20th anniversary of the first human landing on Mars.'

Griffin said that the next step in space exploration would be one step beyond the ISS. 'We will learn from that – how to go beyond – first back to the moon and stay there for a substantial length of time and then on to Mars. In the process, we will build a civilization for tomorrow and after that,' he declared.

Not surprisingly, the private sector too is excited by these ideas. Many believe that investing in space in diverse ways, including developing sustainable commercial lunar business plans, will be a viable option in the future. Google's $30 million Lunar X Prize made big news in September 2007. At a space investment meeting in California in December the same year, Odyssey Moon, a private space business entrepreneur based on the Isle of Man, was the first company to register for the prize. The prize, put simply, is meant to get people up and running in the first private race to the moon, one that will grow the moon business, so to speak. Much of this excitement is fuelled by tantalizing possibilities – of resources, of energy, of base camps for super-planetary trips, of sharpening technological prowess so as to journey to other far-flung planets. The fusion power of Helium-3 is considered an environment-friendly source of energy and there is far from enough of this helium isotope on earth. On the moon, it is another story – solar wind brings with it Helium-3, along with hydrogen, Helium-4, carbon and nitrogen.[8] Also, as Dr Harrison Schmitt, one of the few scientists to have been to the moon, writes in his book *Return to the Moon*, 'embedded continuously in the lunar dust over almost 4 billion years, concentrations (of Helium-3) have reached levels that can legitimately be considered of economic interest'.[9]

The Global Exploration Strategy discussions[10] in the US – brainstorming sessions among the world's experts to answer the whys of returning to the moon – revealed a thematic framework for the future of lunar missions, and the need to test technologies and systems for spaceflight and exploration; everybody was keen on ensuring that future missions, including those to Mars, should be as efficient and productive as possible. Another key theme was economic expansion and the need to extend economic activities beyond the earth. India is already planning a mission to Mars – to understand the Martian atmosphere and processes, its ionosphere, the effect of solar wind, and to search for water and resources.[11]

Anybody Home?

From Armstrong's touchdown to a real base for humans – that is the way lunar missions are blueprinting the coming decades. While sending a prototype, inflatable habitat to Antarctica for a year, NASA recently announced its plans to work on suitable 'housing' for astronauts on the moon.[12] All this and more constitute the grand Constellation Program of the US, which is currently implementing a plan that will get American astronauts back to the moon by 2020 – half a century after Armstrong's 'small step'.

In January 2004 Bush threw NASA a renewed challenge to 'explore space and extend a human presence across our solar system'. To many critics, this is not so much a statement of vision, as a slightly scary question – what will humans stop at? Certainly, NASA is moving diligently towards its goal of a human outpost by the end of the next decade.[13] NASA's Altair capsule will be capable of carrying four astronauts to the moon. According to NASA:

> Altair is a key component in the Constellation Program, which is building the spacecraft, launch vehicles and surface support systems to establish a lunar outpost. This work will provide experience needed to expand human exploration farther into the solar system. NASA currently is seeking input from industry experts and is developing conceptual designs

> for Altair. Between 2009 and 2011, the project plans to build hardware and test concepts. The first crewed flight of the Orion spacecraft aboard an Ares I rocket is scheduled for no later than 2015, when it will fly to the International Space Station. Altair's first landing on the moon with an astronaut crew is planned for no later than 2020.[14]

So we are going moon-camping. But what is the significance of this? In 1966, delivering a talk as part of a national programme series, Sarabhai had said:

> Why does man wish to go to the moon when he has sophisticated instruments including television cameras, which can be sent in spacecraft under command, and can communicate information from millions of miles. It is because nothing that has been developed with the most sophisticated technology so far approaches anywhere near the capability of man who possesses the facility of receiving information simultaneously from a number of channels and to synthesise it to create an image of the environment as a whole. Let us note here that our present day computers and systems for analysis operate only serially i.e., taking one bit of information after another. It is unlikely that man will restrain his image to see, to feel and to listen, himself if he can possibly accomplish all these. I do not expect that the debate on the merit or otherwise of putting man into space would ever be settled. If we are to rely on historical experience, man will surely push ahead with adventures of this type backed by motives which will inevitably be mixed.[15]

It is incredible how true these ideas ring even today, four decades later. It is also true that there really are no rational explanations for human curiosity and the endless search it fuels. Practically speaking too, if countries want to have longer-lasting missions to far-off planets, they would need to break journey somewhere, and the moon may be just the place for that.

Outer space is clearly the next frontier for India, with an unmanned mission to Mars slated for 2015 and a mission to an asteroid planned for soon after. Also, among the most recent announcements from ISRO in 2008 is a plan to send a satellite to the sun. Aditya, the sun satellite, will focus on studying the sun's corona. Towards this end, ISRO is in the process of developing more advanced launchers like the GSLV Mark III, a more powerful rocket than anything in use today. Air-breathing launcher engine technology is also being mastered so as to bring down the cost of the launch.

Where Is the Traffic Police?

The magnitude of lunar activities planned for the coming years, and particularly the possibility of private entrepreneurs entering the arena, scouting the moon like any other attractive business proposition, underscore an urgent need for international guidelines. Without these, there is the worrying possibility that humans, and extravagant societies in particular, will start to over-colonize, overuse and over-exploit the natural resources available – as is inevitable, given our experience on earth. As of now, there are five sets of legal principles embodied within the UN's international law which are relevant to the exploration and use of outer space. These principles are reflected in five different treaties, including the Outer Space Treaty of 1967 and the Moon Agreement of 1979.

That said, countries haven't exactly been falling over each other to have their meanderings in space regulated by international law. Until January 2007, there were just 13 ratifications and four signatures on the Moon Agreement – India happens to be one of the signatories, although it has not ratified the treaty. Interestingly, the US and Russia have not signed the agreement and experts believe this is because all activities governed by these treaties are expected to provide global knowledge and sharing of information, and allow for countries to contribute to human development through their cutting-edge work in outer space. It is widely understood that the lukewarm response to the Moon Agreement is because of its close linkage to the Law of the Sea Convention and the concept of 'common heritage' resources – that such resources, and all activities aimed at the exploration and use of these resources, should be for the overall benefit of humankind. Now that's a problem and everybody knows it.

However, there may not be a choice. Regulation will, in all probability, become an essential factor of lunar and outer space exploration. It may have a tempering influence on several grand and glamorous plans, but that is acceptable. So far, the very fundamental unresolved global issues around legislation of activities in space have not been a serious deterrent to any space-faring nation. But complacency may be a bad idea, given the leaps and bounds with which research is growing. As K.R. Sridhara Murthi of ISRO has noted, this growth may well be the reason for speed and deliverance on resolving global legal issues.[16]

The lure and enthrallment of outer space has been a constant for humans on earth – first as a mystical attraction as with all things unknown, and then more and more as a somewhat enigmatic but approachable entity. In fact, to people like Schmitt, the moon is not just approachable, it is also a great business proposition. In his *Return to the Moon*, he asks, rather dramatically:

> Will we begin the settlement of the solar system and provide for a 'new birth of freedom' beyond the Earth? Placing the doubts behind us, let's continue the millennial adventure that began in 1969 and go where we have gone before, and then beyond. We owe the future of humankind another walk on the moon!

Grand challenges fuel great discoveries, but the concept of a bustling space colony, thrilling as it may sound, generates considerable disquiet. Limiting the use of spatial resources by people and governments, through legislation, makes eminent sense, before earthly beings with all their baggage of life and living spill out into the no-longer baffling or unexplained darkness of outer space.

Epilogue

In a depleting global resource environment, the earth, moon and Mars have really to be viewed as one large commercial complex.

Dr A.P.J. Abdul Kalam

In this and the next decade of the new millennium, the relationship between the earth and outer space, between humans and planetary bodies, will be rewritten. But it is hard to imagine a future devoid of the mysteries of the solar system, and of outer space. The idea of an outer space dotted with human settlements and with rockets zipping back and forth, is not particularly attractive either.

Remember the Jetsons? The popular cartoon series on television had its main protagonist George Jetson working for Mr Cosmo Spacely, and going to work in a vehicle that looked like a flying saucer. The Jetsons obviously lived somewhere in outer space and the creative concept was first born in the early 1960s – the story seemed to hint that such space-life would be normal a hundred years from then. Later, it seemed more likely that it would happen at the turn of the twenty-first century!

Is there a touch of rashness in this speedboat approach? There is certainly a lot of one-way traffic expected to crash on the moon since the earth's nearest neighbour seems to have come back in vogue as a hot destination.[1] Even as the missions take off and the world gets busy once more, courting the moon, planetary scientists are worried that the surfeit may cause a virtual traffic jam on the lunar highway. That apart, pollution could become a serious issue.

These voices of concern rose at the fifty-eighth International Astronautical Congress held in Hyderabad, where scientists from 40 countries converged. A primary cause for worry was the fact that all the moon-orbiting satellites would crash one after another onto its surface. Some of these crashes would be planned and deliberate, of course – in search of water at the lunar poles. Other satellites would simply be abandoned once fuel ran out and the mission's life sputtered out.

The first to end up on the moon would be Japan's 3,000-kg Kaguya, a satellite with a year's life and the heaviest to go to the moon after the Apollo missions. Then there's the Chinese Chang'e mission, with a 1,900-kg satellite and a life of one year. Chandrayaan-1, a two-year mission, weighs 590 kg in the lunar orbit; the 29-kg moon probe is to hurtle towards the lunar surface soon after the satellite reaches the moon. The Lunar Reconnaissance Orbiter of the US, scheduled for the latter part of 2008, is about 1,000 kg, with a life of one year.

Some of these satellites release gases that could alter the moon's environment, although the metallic parts of the spacecrafts are likely to remain more or less intact since there is very little erosion on the moon. Several hundred kilograms of human waste are already piled up on the surface of the moon, left over from the Apollo missions, and more will be added in the next two to three years, contaminating the lunar surface with materials alien to it. Experts such as Dr Bernard Foing, executive director, ILEWG, the Netherlands, are clear that something needs to be done. Foing is known to have said that the 'contamination of the moon is a problem that needs urgent attention'.[2] It is certainly a 'genuine concern' for many others too, such as Dr Carol J. Russo, Deputy Director, Exploration Technology, NASA Ames Research Center in the US who adds that 'we need to work towards not leaving junk on the moon surface but to make sure all that is left there is used to make the moon surface habitable in the future'.

There is in place an international law that governs such activities. The United Nations Agreement Governing the Activities of States on the Moon and Other Celestial Bodies of 1979, popularly – and understandably – called the Moon Treaty or the moon agreement, is an international agreement that turns jurisdiction on all heavenly bodies over to the larger international community. Thus, all activities in outer space must conform to international law. As always, the gap between the crafting and implementation of laws is significant, and the treaty has been a non-starter in more ways than one. Not ratified by any nation that engages in manned space exploration, it has negligible impact on regulating global activities and actual spaceflight. Put simply, the Moon Treaty has failed.

Several experts and nations are pressing for greater collaboration and coordination between governments. This is critical if we are to ensure that the lunar environment remains the way it is – unpolluted and clean. Dr Roger-Maurice Bonnet, a solar physicist and President, COSPAR in Paris, worries that the moon is going to be 'spoilt' since human society is 'planning to do things we don't even do on Earth'. There is a need to create a 'scientific conservation area' on the moon, he says, that will remain unspoilt.

That none of the planned space missions have announced any exit plans for their spacecrafts is something to worry about. China's chief lunar exploration scientist Wu Ji says there is a need to create a specific location on the moon that can act as a 'wastebasket' where satellites whose missions end can be dropped. Foing suggests that it might be easier to put satellites 'to sleep' in their orbits rather than letting them crash on the moon. Bonnet feels the 'moon will be destroyed sooner than later if lunar pollution is not attended to in right earnest. You need an exit policy from the moon'. According to him, it took 27 years for the lunar atmosphere (or whatever there is of it) to stabilize after the Apollo missions.

If human missions were to spoil the untouched historical repository of the evolution of the solar system, it would be a real pity. At the same time, there is no doubt that the magic of exploration is a heady potion that must be sipped by all. Some prudence and some rational planning is all it would take to ensure that we strike the right balance. Every mission that goes up has to have clear and necessary objectives, and should serve the greater human good. Once that justification is alive, the checks and balances, the avoidance of traffic jams on the lunar highway and ways to avoid garbage piles on the moon can be attended to. As of now, it seems that growing the cadre of people and countries that would undertake only prudent space travel is the sole sustainable option. If that can be done, it would all be worthwhile.

Annexure 1

Missions Galore

Interview with Dr G. Madhavan Nair, Chairman, ISRO

by Pallava Bagla

Q. So tell me a little about this man in space proposal which ISRO is making, which I am told you are spearheading?

A. Well, you know, in space technology we have achieved a level of maturity. We have the capability to build satellites of any type, whether it is for earth observation, communication, or the science missions, and also place them in orbit with our own launch vehicle. So this is actually a totally self-contained and indigenous technology on par with any of the developed countries'. Naturally, the scientific community, especially the ISRO community, will be looking forward to 'what next'. I think the scientific exploration of the planets, the solar system, the galaxies – these have been in the minds of people for ages. Even when ISRO started, we were investing in such programmes. At least about 10 per cent of our budget has been utilized for such explorations. In the early days we had small payloads within the satellites and now there is a full-fledged system. Chandrayaan is a complete scientific satellite, and also Astrosat, again a sort of observation platform in space which can look at the galaxies.

Logically, such outer space exploration can be done in a much more meaningful manner without the interference of the atmosphere. One can go in for instrumented systems, but at the same time human intelligence and capacity coupled with powerful instruments can provide higher dividends. As far as humankind is concerned, yes, we have a lot of problems to solve on planet earth. But at the same time, the next frontier is going to be outer space. For various purposes, whether scientific observations, manufacturing, power generation or even from the security angle, outer space is going to be the most important location. So our presence there will be a logical extension of any space research activity we undertake. And if we say that we are going to maintain a sort of leadership in space research, naturally the exploration of space, outer space, with human beings, will become part of that. So we need to see what must be done in case we undertake a manned mission.

I think I have myself gone on record earlier, that the cost of such missions is very high, and of course instruments and robots can do many of the jobs we are planning to undertake. One has to weigh the pros and cons of an automated mission vis-à-vis a manned mission. This is also a sharpening in terms of what new technology has to be developed for the manned mission, and whether it is within our reach; the new facilities we need, the type of advanced R&D we need – this has to be assessed. We are taking that up. Our centres – especially the Vikram Sarabhai Space Centre in Thiruvananthapuram, ISRO Satellite Centre, Bangalore, Space Applications Centre, Ahmedabad and Satish Dhawan Space Centre, Sriharikota, all joined together to make an assessment. This assessment is what was presented to the peer review committee and we have projected all the pros and cons related to this issue. We had a very lively discussion: senior-most scientists from the country, not only within ISRO, but outside ISRO, institutions, industries and so on, they were all present. In the end, the convergence was yes, the technology is within our reach.

Second, they also endorsed the view that human presence in outer space is the next step in the space research programme. But of course, one thing they have said is that when we are talking about a manned mission, safety is very, very important. It is not only the success of the launch, but the break in the orbit, bringing them back safely, and even in the course of the launch if anything goes wrong, how to have something like an ejection system and so on, you know; how human safety can be ensured, that has to be studied in depth. Also, we have to clearly identify the new facilities and technology required. We have taken this suggestion into account and asked our teams to make an in-depth study and come up with a project report as quickly as possible.

Q. Why this sudden turn? Some months ago I remember you told me that we have no plans to put man in space. What has suddenly changed?

A. Nothing has happened specifically as a transition. It is a continuous process. Actually, even earlier, the studies had been going on in our centres. But it was premature to talk about it at that time. I was also a bit worried about the cost impact. You know, the whole planned budget in the Tenth Plan was around Rs 13,500 crore. Out of this if I have to spend Rs 10,000 crore for this, I am left with nothing to meet societal needs. So that's why I was a bit hesitant. But now there are two things – one, we have a sharper assessment of the technologies needed, whether we have the capability, whether we can harness this within a reasonable time – that confidence has come from our internal studies. Second, the country's economy is growing very fast and maybe in a few years' time from now this Rs 10,000 crore will be affordable for the country. With these criteria put together, we thought we should definitely make a proposal to the government and get it cleared. But at the same time, our thrust on societal applications will continue.

Q. That is what ISRO has always done?

A. That continues with full vigour and we are not going to dilute any of those processes. In fact, by the next plan (Eleventh Plan) we will have a very powerful observation system in terms of radar imaging and hyper spectral imaging, even the capability to observe from the geo-stationary platform. When disasters occur, we can have a continuous watch on that place. On the communication side also, we are going to provide newer capabilities; DTH is already catching up fast. We are going to have K_a-band transmission by which we will be able to give large bandwidth connectivity to remote parts of the country. Such societal applications are going on full steam. But parallelly we will try to see to technology development related to a manned mission, and if the government decides, in about eight years after the decision we should be able to have a manned mission.

Q. This will be in a low earth orbit?

A. Yes, it will be somewhere about 300–600 km orbit. Basically, you know, it is not just launching a man into space but the technology thrust that it is going to provide, not only for ISRO but for the entire country. It is going to be a national project where multiple institutions will be participating. The type of – especially human – behaviour under zero gravity conditions, and the type of radiation protection system we need for these, and the reliability of the system will be studied. All this would be a technology driver. Even if you take other developed countries, many of the commercial technologies have been spinoffs from space. Even today, and here too, it happens that there are many areas in which space technology has been adopted for practical application. Once there is a manned mission, a higher level of technology will come into the country. And that technology base will be a national asset.

Q. 'Come into the country', you said – are we looking at a joint mission with somebody?

A. No, no, no. I should say we will have, the country will have, a higher level of technology.

Q. But tell me what transpired between March 2006 and November 2006? I am very intrigued. Between March and November there was a complete U-turn. You know, in March, when we were talking about NASA, NASA was saying that they can help us put a man in space?

A. NASA…we discussed it with them. NASA's suggestion was that they train astronauts in their facility. Then we asked specifically, can they accommodate an astronaut in their space shuttle? Unfortunately there is no slot available. All the launches scheduled from now until 2010 or so, when the shuttle will be de-commissioned, they are all committed for orbiting the space station. Because of this, we didn't pursue the topic. They are talking about training only. But here we have to think of establishing the training facility ourselves, gearing up on our own, what you call aviation medicine-related activities and then equip ourselves for carrying out this task.

Q. So NASA was not keen on putting an Indian in space. They were only interested in training?

A. They said if you are interested we will train your astronauts. That's the suggestion they gave.

Q. But is that a suggestion that you are still willing to accept?

A. Well, it depends. Actually, we will try to give the first priority to evolving a highly reliable launch system. Maybe a derivative of GSLV Mark II or Mark III. Both vehicles are capable of carrying such manned missions. Technology development is of prime importance. The facilities for training astronauts, I think the country can establish. The infrastructure exists, from training pilots in high gravitational force environment and so on. Perhaps we should establish our own facility for training. These are national investments, I would say. Even if you don't send a man into space, these facilities and the type of studies and the results will go a long way.

Q. But do you think the country can afford to have this mission?

A. That's what I told you, today we are talking about a growth rate of about 8 per cent at least. With that kind of economic growth rate, even if the space programme budget can be doubled, yes, these numbers come within our reach.

Q. Now what did Prime Minister Manmohan Singh say – before you had this meeting with the scientists, what you call the peer review, you made a presentation to the prime minister and brought him on board?

A. Yes, I have highlighted that we are doing these studies. Then he himself asked this question, you have to justify the pros and cons and come out with the detailed study. He supported our view and sought wider endorsement from the national scientist forum. So, I would say that he has listened very patiently to our presentation.

Q. And also given a nod?

A. No, that I cannot say, yes or no.

Q. You took it forward. If he had said no, then you would not have taken it forward. It is tacit approval.

A. No, actually, any such approval will have to go through detailed assessment. You know, for Chandrayaan, we spent nearly three years making a detailed study report, getting it vetted by various agencies, etc. Though not three years, it will take at least a year before we can converge on this and I can take a concrete proposal to the PM for his approval.

Q. Again I come back to the issue of costs. How do you justify putting a man in space, in a country where there are 300 million poor, one-third of the population is below the poverty line, 40–50 years after the US put a man in space, or Russia, and now China? How do you justify that to a country?

A. You look back: in the seventies we were asked the same question, why should we spend money on space? I think we had not even scratched the surface. We only had sounding rockets. But we boldly said that we would build a satellite, we would build our launch vehicles and we would make it relevant for society. In fact, if we look back we have spent maybe close to, let us say, about Rs 10,000 crore on the space programme up to now. Our space assets, 16 satellites up in the air, that itself is about Rs 6,000 crore. The benefits brought to society in terms of natural resource development, communications services, all those would cost at least twice the amount spent on the space programme.

Q. So for every rupee spent you are getting two rupees back, is that what you are telling me?

A. This is what we are giving back to the country – two rupees at least.

Q. Is this based on a study or is this based on mere projections?

A. In fact, we didn't want to make a projection of this kind at all. There is an Institute of Economics at Chennai, they have made a formal study that reports the conclusion. If you take a simple example, we are providing technology to Doordarshan channels, the communication channels. If you want to hire the capacity of transponders of any foreign country, that costs at least three times more than what we are spending. The other day I was trying to give an example – of the national drinking water mission where the success rate of the yield in water has gone from 40 per cent to around 80 per cent, but if you count the number of wells which have been dug, that itself will be something like Rs 500 crore.The study uses several such examples.

Q. This is interesting because one has heard of these things as anecdotes but now you say there is a full study?

A. Right, so similarly, the entire technology base in the country is going to go up. That is the first indirect benefit you are going to get. And we are going to get a highly talented human resource pool, trained and available for handling any programmes. Then the third, of course, our capability to access space independent of others. See, many of the people, when it comes to a critical use, they will not share such capacity with anybody else. As a nation which is moving forward, which is trying to be at the forefront, we should have this capability with us.

Q. That's why you are targeting putting a man in space by 2020?

A. 2020 is too far; I want to do it much before that!

Q. When do you want to do it?

A. As I told you, about eight years is the timeframe which we are talking about after the government's approval.

Q. That means 2015. That is man in space. And when do you want to put a man on the moon?

A. That is a little far away, because to reach the near earth orbit we need something like 11 km/sec as a velocity equivalent, whereas if you want to go to the moon and come back, it will be almost double. So that means much larger launch vehicles are required and that development will take time. Maybe after the initial mission of a man in space, six to seven years. But the costs are going to be much higher.

Q. So that cannot be achieved by GSLV Mark III?

A. No, because it's a capsule, very small, and it cannot (take a man to the moon).

Q. So are we talking of a science mission initially, or a technology demonstration?

A. Initially it will be a technology demonstration, and then based on that, it can go to the next step of being a science mission.

Q. Because repeatedly people have told me, men in space don't do good science?

A. (Sighs deeply) Well, I think, on this, I cannot give you a firm answer. As one of the scientists said, suppose you scan a portion of the sky from one end to the other, and then try to look for an object, it will take a few days. But if a man is there, you know he has got a very wide perspective and he knows what he should look for. That way, the whole process will get accelerated. The power of the human brain is not easily replaceable by a machine, and the speed at which instruments can do this job is nowhere near that of human beings. This is the view of all scientists.

Q. No. When you put a person in a capsule, the space he has for doing anything is very restricted. Because of that restriction he is unable to conduct good science. And this has come straight from Prof. Yashpal, he himself told me, and I believe he told you in the peer review meeting also. He was very clear about that.

A. He did put that point across, so did another scientist. Let me give you a simple example. You look at this room. Let's say you want to study more about that fan or that projection system, then your eyes directly point towards it. You won't look at other things. Whereas instruments mean we start scanning from one end to the other, then we try to decipher what is what. Human observation is more efficient. But yes, human adaptation in space is a fairly difficult task, it requires extensive training, and maybe sometimes a scientist may not be fully attuned to that environment. If you look at the space station, they have established it fully and that platform will be available and there may be the possibility of tie-ups at a later date. Our own access to the platforms will help us do the job better.

Q. We are talking about access because we have never been part of the International Space Station (ISS).

A. No, we have not been part of that.

Q. So are we now talking of joining the International Space Station?

A. This has to come from the user community. As I told you, the scientists have to come with specific requests – what they want to do with such platforms. Only then can we think of having a dialogue with others.

Q. But you are open to the suggestion that India can participate in using the ISS?

A. Yes, if there is a specific need, definitely one can.

Q. Do the other countries want India on board?

A. Well, the US is the main partner, and now you know Indo–US relations, they are improving quite a bit. So we'll have to wait and see.

Q. The same thing was considered with the International Thermonuclear Experimental Reactor (ITER). Now India is part of ITER. We were not part of the ITER experiments?

A. See for example, the US is participating in our Chandrayaan mission. They have sent two instruments to be flown piggyback, so that's a positive step towards improving our relationship.

Q. So you are saying that India may not be averse or as allergic as it was towards ISS? We could be participating in that experiment?

A. No, as part of international cooperation, such things (can happen) and there is no reason why we should be rejecting it.

Q. But are we also planning to put a space station kind of object in space?

A. At the moment such things are not on our horizon. But you know, some of the scientists are talking about the possibility of building a platform which can be used for generating power and then beaming it back to the earth and so on through a connection.

Q. But are these Indian scientists?

A. Yes, even foreign scientists have got some concepts on that. But there are a lot of technological breakthroughs one has to make to be successful. See, first of all the efficiency of the solar panels will have to improve to at least about 50 per cent. Then, a proper system of transmitting this power from the space to the ground without it becoming a safety hazard. Such things have to be solved, only then you include it in a proposal. But there are many things which were dreamt of 20 years back and have become a reality today. So we cannot rule out such possibilities for the future.

Q. So you want to be there, prepared?

A. Naturally.

Q. And are we looking only at the moon or beyond also?

A. There are some scientists who are interested in studying Mars-related issues, there are others who are interested in solar weather, and in the sun–earth coupling. So these things will extend our space-bound travel beyond the moon.

Q. So we are looking at the entire gamut of the solar system today?

A. Planet earth is influenced by the solar system. So the issues related to the solar system and its coupling with the ionosphere, atmosphere, are very interesting studies.

Q. So ISRO is looking at the moon and beyond now?

A. No, ISRO is open to any such suggestion from the scientific community, and based on the scientific community's interest we will be conceiving the mission.

Q. But you drive that?

A. Well, we act as a catalyst.

Q. Now, coming to GSLV. We have started talking of man in space even before we have our own indigenous vehicle in place. The critical component is still imported. The cryogenic engine?

A. No, the cryogenic engine is getting ready, the flight stage for the next mission is getting ready. So we have a full handle on that.

Q. Still, it has never been tested?

A. I told you, there is a full stage. We have something like the ground test.

Q. And when do you hope to put it on the GSLV and have it up?

A. Maybe 2008. The stage is getting built. It is in an advanced stage of preparation.

Q. Well, the proof of the pudding lies in the eating.

A. Naturally. It is a very, very complex system. We have done a lot of innovations, so we are happy that all these new elements have performed very satisfactorily, and even compared to the Russian stage, we get an advantage of nearly 200 kg in terms of payload.

Q. And the same thing will be used on GSLV Mark III?

A. No, Mark III has got a higher thrust level engine, which is getting developed. This current engine is about 7.5 ton. The Mark III engine will have a 20-ton thrust.

Q. So GSLV Mark III is going to be a 100 per cent indigenous vehicle?

A. Yes.

Q. In fact, that was one criticism that came your way which was that soon after the GSLV failure, you made these grandiose announcements about putting a man in space. How do you react to that?

A. No, it was only a coincidence. It had nothing to do with the failure, because our programmes are slated well in advance. Even for a launch, preparations start about two years before. So unfortunately yes, this launch of GSLV had a problem because of one of those manufacturing defects which went unnoticed. But at the same time it gave us confidence, there is no design flaw in any of the subsystems of the vehicle, so that we are not really apprehensive about the future of the launch and so on. We have taken all appropriate action for reinspection, retesting, so that the chances of such an event occurring will be very low for future missions.

Q. The first time there was a lead nugget or something in the pipe, so the nozzle was thin. This time the nozzle was wider, more open?

A. It was a different component. That one was on a plumb line, this one was on a regulator. It had a larger opening than desired. This is very challenging as the type of performance demanded is very, very exact. So one cannot compromise at any stage of preparation of the subsystem. Of course this is known to us, we have been cautious, we have got our systems in place.

But still you know, when we work with any system, when you start looking at the launch vehicle, it is nightmarish. There are, at least we know, more than 100 subsystems without any redundancy. They have to work, within a performance band I will say, within a couple of per cent of the nominal. So that is a really tall order in terms of any technology, and here again, many things are not testable on the ground. It will get exercised only during flight time. So all the ingenuity, proper engineering, manufacturing, testing, calibrations if required, and so on, needs to be perfected.

Even if you take the last ten years' history of launch vehicles all over the world, even after stabilization, they have had failures of about 10 per cent. Globally out of 4,000 or so launches till date reported failures are around 390.

Q. So it is in the acceptable band just now, ISRO's failure rate?

A. I will say, well within the acceptable band and also the fact that since it is not related to any other design or mission-related problem, we have full confidence.

Q. I would disagree with you there. You see, if you can stop a vehicle when there is not enough thrust, which is what happened in the first GSLV launch, you stopped it beautifully, I was there at Sriharikota then. Now in this instance when there was extra thrust, there should have been a system in place to have stopped it.

A. You can, but at the same time, you know, we have debated this, whether we should have upper and lower limits for these. But these engines being a sort of hydraulic control system, you cannot precisely predict their transient behaviour. So if the transient exceeds the nominal value it would put a hold under certain circumstances and the real mission would be stopped, aborted. So that's why we didn't exercise that option.

But now we have activated a team to study whether we can evolve an appropriate criterion for the upper limit also. Perhaps that could be one extra caution, but at the same time the chances of a normal mission being aborted also have to be considered. So it's not an easy thing, we debate considerably. There are possibilities of putting a redundant system working in parallel. But there are two modes of failure – one would be lower performance, and another would be higher performance. So two elements put together cannot take such a decision. Then you know invariably we will go for a third element, or maybe fourth to make doubly sure. Once you put four elements, then the mass and everything goes up. One has to really weigh and take a calculated decision on this. So we go through this process. But launch vehicle design is not a simple job. People really burn thousands of hours, sitting around the table discussing, debating on such issues and to draw a line where it needs to be normalized.

Q. So what were the lessons we learned from that dramatic GSLV failure?

A. One, we have to be extra vigilant with respect to acceptance procedures, especially when we are spreading out the subsystem manufacture to various industries in the country, because all industries may not be equally tuned for carrying out this high quality work. And we have to have more number of streams.

Q. Any other?

A. That is the main lesson which comes out of this.

Q. Now if we talk of the satellite recovery experiment (SRE), it is really the first step towards having a man in space because mastering re-entry is critical to a mission?

A. Right. Basically learning some fundamentals related to re-entry. Breaking the orbit, bringing the capsule safely back home and controlling it through the extreme regimes of aerodynamic heating and so on, those things got validated in this experiment.

Q. We also tested, what did we use for re-entry heat control?

A. A few thermo protection materials like part of phenolic materials are sometimes used, then silica tile, which is used for high heat insulation, then some ceramic coatings. So it's a combination of systems which is evolved for these particular heat conditions.

Q. So these are all locally made? No imported technology in this?

A. Yes.

Q. And this is the first time we tested it?

A. That's right, so this was a very tough mission, because 7.5 km/sec in orbital velocity, when it touches down it can have only about 15 m/sec. So the initial breaking was done by the aerodynamic shape itself and later with a set of parachutes which are used for this. And then precisely calculating the re-entry trajectory to make it fall into the location, the sea – so these are all real challenges.

Q. *It was a sea drop?*

A. Yes.

Q. *Are you catching it in midair or it's going to drop?*

A. No, no, it uses a floatation system, so as it impacts it floats.

Q. *And is this what we are planning to do for the man in space also?*

A. Normally a manned mission will also have a similar capability.

Q. *Is this capsule sufficient to carry humans?*

A. No, no, this is only a 500-kg capsule, the manned capsule will be 3–4 ton.

Q. *So a several-fold increase has to be made?*

A. Yes, that is why I said that technology development itself will keep us busy for the next four or five years.

Q. *But we are talking of a capsule system, not a shuttle system?*

A. Now everybody recognizes the shuttle system is not an optimal configuration. Even NASA is going back to the Apollo module to be refurbished for their next generation vehicle.

Q. *So we are going to go the capsule route?*

A. Yes.

Q. *And are we looking at collaborations from other countries in this, or this will be 100 per cent Indian?*

A. We have not explored that possibility yet. I think if at all the technology is available, it is only with the Russians and the US, so let us see how things develop.

Q. Even China?

A. In China, I think, it is more like Russian technology.

Q. It is possibly Soyuz, some say, but the Chinese themselves say it is indigenous.

A. Perhaps, indigenous adoption is also there in the technology.

Q. So is our capsule also looking very similar to Soyuz?

A. No. Shape-wise I think, it has to be an aerodynamic shape, you know that is decided by physical parameters. It has nothing to do with the Chinese or Russian designs.

You take the Apollo capsule as well as what the Russians are using, there is a lot of similarity. Diameters may vary slightly or cone angles may vary, but we will go through our own studies. You know there is no point in trying to copy somebody else's design. We have to conduct wind-tunnel tests and then try to arrive at optimal parameters, and then of course, we will build on our own.

Q. So what will be different in this from others? Do I remember Prof. Yashpal saying, don't copy?

A. Yes, see, for example in the spacecraft recovery experiment the type of insulation which we used is a very sophisticated material, which neither Soyuz nor Space Shuttle have used. Of course Space Shuttle has got almost similar material, silicon foam, but not of the same form, because it was developed almost 20 years back. We have taken advantage of the latest technology and the type of foam we have developed is much better than any of these materials. So the proof will come out of these tests. So that's how, you know, like rockets, the shape is the same, and the principle, but how you evolve as a system, that calls for a lot of engineering.

Q. Now you are talking of 2014 and 2020 – is it largely because that is what President Dr Kalam talked about, making India into a developed country? As he said, a developed country without the capability of putting man in space is no good. Is that something that has influenced ISRO?

A. Of course, Dr Abdul Kalam has really been an inspiration for all of us. Whenever he comes to our establishment or meets our scientists, he keeps on talking about the far future, not about what has to be done tomorrow. So he talks about Mars exploration.

Q. Or he has ordered ISRO to do this?

A. Yes, that is also one of the inputs, but not a direct input.

Q. Not a direct input. Obviously he is not an executive president. And how will you avoid not copying, because that is something which has bothered several people?

A. See, our scientists are very innovative. They are able to bring out good ideas. Even for solving a problem you will have multiple solutions in front of you. Making the right choice, and then trying to arrive at an optimum solution, that is the way technology grows. So we will be always looking at that direction. If you want to simply copy the Soyuz module, maybe even if you go to the literature, about 80 per cent of the information will be available. But that is not sufficient. We believe that it has to be from scratch.

Q. There is this talk of a little regional race which is happening between India and China. India announced the moon mission, China came forward, they launched man in space, and now we are putting a man, possibly on the moon, ahead of them?

A. I don't think we are trying to get into any race with any other country. It will be purely driven by our own need, and our capability, and maybe the resources we can spare.

Q. So you are saying there is no race?

A. Yes. There is nothing, we are not looking at others and then trying to see what we should be doing. If our scientists believe that the manned mission is essential and if the country can afford it, yes, we will embark on it.

Q. *Even if you are beaten by China you are not particularly worried?*

A. No, they have their own priorities; they have already gone ahead and demonstrated that. But at the same time in some of the areas they are not as advanced as we are. So we have to build up on our strength, rather than trying to be competitive with somebody else or copying somebody else.

Q. *Continuously I am told that collaboration is the next big thing. Countries putting their heads together, joining together, launching missions together. Is that something which is setting in at ISRO because you have the confidence now?*

A. We have international collaboration very much on our minds. A recent example is called the Megha Tropiques project. This is a satellite for studying tropical atmospheric phenomena. France is contributing major instruments and we are launching it. Then we have cooperation with Russia in what is called the Youth satellite: when our president was there he declared that some of our universities and Moscow University will be cooperating in building a satellite for science. Then of course the moon mission is a very good example of this. So we are on that path, we are open and with many other complex missions we will try to bring in international cooperation.

Q. *How hopeful are you of getting a nod from the government for this man in space endeavour/project?*

A. I cannot predict at this stage. Let me go and talk to the government with a formal report.

Q. *Rs 16,000 crore is what you mentioned.*

A. No, that number has to be sharpened further.

Q. *It's varying between Rs 10,000 and Rs 16,000 crore. There is a lot of difference.*

A. No, see, I want to do a thorough exercise as to what facility is already available in the country, what needs to be added, what new technology has to be developed, and if that is the case, what is the cost – so all this has to be estimated precisely. Since that outline exists now, we should be able to build upon that, and then in a few months from now we should be able to submit a project report.

Q. Then you are confident that ISRO can undertake such an arduous mission?

A. That, of course. Yes, we can.

Q. Despite the GSLV failure, you are still very confident?

A. Yes, from failures we learn much more than from success.

Q. Okay, now shifting track, I was in China recently. I interviewed Dr Wu Ji, the chief lunar scientist for China, who is doing the lunar payloads for Chang'e. He mentioned to me that they are very keen on having a joint mission with India very soon, even though it is at the level of scientist to scientist. Dr J.N. Goswami from Physical Research Laboratory has told me that it would be opportune to do that, but has there been any discussion between China and India to collaborate on space technologies?

A. On some fronts, we have already established some contacts. For example, our remote sensing images from IRS satellite, I think it is being received in China and being utilized for their purposes. In the science mission as you rightly said, recently there was a conference in China, the COSPAR meeting, and in that they have discussed many things. When the proposals come we will examine them, we are open to that. But at the moment we don't have any concrete proposal with us.

Q. But are you hopeful that the two countries can work together?

A. See, it depends on the technology front. I don't think immediately we may get into any cooperation because most countries would like to do it themselves. At the same time, in the science and application side there is enough scope for joining hands. It depends on the interest of the scientific groups.

Q. If you see the recent joint declaration between Manmohan Singh and Hu Jintao, that has a whole paragraph on space. Now that is something new, the last declaration never had that.

A. So we have to build on that. You see, normally at the highest level they make the principle and the details have to be worked out at the working level. This is the first time I think we are having such an open dialogue with China. So we have to put across some other points, and this is the convergence which has come. This identifies the broad areas where we can cooperate.

Q. I was told by the foreign secretary that there is discussion of reciprocal swapping of satellites and also joint missions.

A. There are a number of possibilities. You see, just like in the moon mission we are going together with Europe and America, maybe in some other science mission it may be with China and Japan perhaps. So there are possibilities. See, these contacts have to be established at the working level between scientific groups. Then the other things will follow. First of all you know there may be a capability for building an instrument here or there, and then they should come to an understanding that we will share data and have a common analysis. So that has to evolve.

Q. But it is fascinating. China wanted to launch the INSAT satellites using the Long March rockets. They said it was very cost effective. Possibly it is, and we didn't take it up at that point?

A. Actually we had a problem about reliability and such issues, and so before we make up our mind about a particular launch configuration we will look at not only the cost but what is the track record of the missions and so on. Especially when you are buying launch services, it goes by a competitive bidding process. So we will put in a request for a proposal and all those who have the capability will send a quote. Then of course quality and the price, that's what decides finally. That is another area which is identified, but let us see.

Q. So are we looking at any future launches on Chinese rockets?

A. At the moment I don't think there is any consideration but in the future it can evolve. Because it depends, you know, on where the gap is. If we have at least up to 2.5 ton, we can do that. When we go to the 5-ton capability, yes, but we also will have the GSLV Mark III available soon. But here it is not only getting access to the launch services but all the other infrastructural support required. You take a satellite launch base, whether it has the preparation facilities and so on has to be examined. We will try to assess this.

Q. And where is our cooperation with the US heading now? Let us concentrate on the US side because with the civilian nuclear deal almost done, high technology cooperation is likely to happen.

A. In fact, perhaps on our path of Next Steps in Strategic Partnership (NSSP) what we have achieved with Chandrayaan mission is a significant step forward. We now have the understanding of the two American payloads with Chandrayaan, and that has gone through the entire process of the Technology Assistance Agreement (TAA) and so on. At least it's set a benchmark on how you can proceed on the cooperation. Then in Gagan, the Raytheon company from the US has contributed to the establishment of the ground stations here. I think it is moving forward.

Q. But in India's manned programme, are we hoping to cooperate with the US?

A. Unless we have clarity we cannot go and have a dialogue with somebody else, you know. We ourselves have to get clear, then we will consider having a discussion with others.

Q. But you have not had a dialogue at all with them on this new vehicle they are developing?

A. No. Actually, I think they are very categorical: as far as access to space is concerned, it will be their own development. I think it is not only the US, no country is expecting cooperation on rocket technology from others.

Q. And the same thing applies here – that is why we are getting our own independent capability?

A. Yes.

Q. I am told you are absolutely sure that India has to have its own launch capability?

A. (Laughs) Yes, well, I think if you take it as an astronaut's prestige, this is my personal opinion, nothing official, I would like to have our own flag flying on our vehicle. Also, twenty years from now, when space travel is likely to become mundane like airline travel today, we don't want to be buying travel tickets on other people's space vehicles.

Q. So you are essentially saying strength begets strength.

A. Yes.

Q. When you have the capability, others will be willing to join hands; if you don't have it, nobody will be willing.

A. Absolutely. Our technology is recognized. We are getting orders from European companies for building satellites here.

Q. So in this sector would you say India is a developed country?

A. A very, very developed country!

Q. Now shifting to manpower – that is something which has bothered ISRO. There have been a lot of people who have quit ISRO for the private sector. Is that something which worries you, and what are the figures?

A. At one time we were really badly affected. That is when these IT companies started moving here. Maybe one or two years we had, at a younger level, even 20 per cent depletion, but...

Q. Which years were these?

A. I think it must be 1999–2000. But afterwards, we organized and stabilized ISRO. I think we can boost our salary, we are hardly about 7 per cent less, but the problem is to get the people. We are not able to really attract the highest strata who are coming out of IITs or leading institutions, mainly because of the wage disparity. The IT and service industries offer salaries that are way higher compared to what the government is offering. So this naturally leads to only the next best coming to us. But we go through a very rigorous process of selection, having a written test and interview and so on. The yield is poor but the quality is good. Last year's experience I wish to tell you – starting with a raw stock of about 75,000 people, students who have come out at graduation level, at the end of the written test, only 1,500 qualified. Out of 1,500, we could select hardly about 150. So that problem has to be tackled.

Q. And how many have you retained?

A. If they are into a stream for two years, at that point nobody leaves. Because they love the job and are now attracted by the technological challenges and also the working environment. But to attract at entry level we are facing serious problems.

Q. And how are you hoping to correct that?

A. Well, we are trying to publicize. The wage that they advertise in the private companies vis-à-vis the government, it is not one to one. In private companies they add up all the perks and services and then they put in a number. But here housing, medical, food, transport, all these are heavily subsidized. If you put those numbers, they more or less match. We are trying to give a message, but still you know, the youngster who is just coming out of college, all this logic doesn't count.

Q. Is this man in space business also a way of attracting talent? Because you are giving a challenge?

A. Certainly that will be true. I tell you there are still many youngsters who really enjoy the technology challenges. They just don't want to stick in front of the computer and do very routine jobs. In fact, in the last two years, a few have come back to ISRO from the industry though not in very large numbers. But we have to improve their remuneration, something needs to be done about that.

Q. Also they say people in ISRO are getting bored, doing repetitive things, making a satellite, making a rocket.

A. It's not really true. Repetitive production, we are farming out to the industry. We concentrate on development activities. Once the system is developed, the production is entrusted to industries. At the same time, when it comes to the mission, our people, yes, they have to be there. But at the same time it is very thrilling and challenging for them. They enjoy every mission, though it is the same PSLV, same GSLV launch. Yes, the excitement is as good as, as challenging as the first launch.

Q. So when do you really hope to farm out as you said, all of this repetitive work, because essentially making satellites, making PSLV, making GSLV, once you have mastered the technology, when do you really hope to pass this technology on to the industry completely?

A. There are two types of technology. One is very critical, sensitive technology that we do within our laboratories. But if it is a routine production job, almost at the first instance we hand it over to the industry. Today the examples are, about 60 per cent of our jobs which we require for the launch vehicle are coming from industry. And about 40 per cent for the satellite, that is also coming from industry. So this is a good starting point and as time goes on, we will increase this fraction.

Q. When do you hope to finish this? Say Dr Madhavan Nair sets up a company and completely commercializes PSLV and GSLV, something like that. When can it be expected?

A. Unfortunately our private sector has not come up to that level yet because it involves investing in human resources and retaining them for a longer period. Our private sector has not really caught on. But maybe when they taste blood, they will come up with some good solutions.

Q. Have you tried doing it?

A. Yes, we are open. For example, right now we have put up a request for a proposal for integrating the satellites in our facility, because the facility cost is very high. So we are asking industries to come with proposals. If they come with good proposals, we will accept them.

Q. So they will do the complete integration? And when is that likely?

A. It depends. In fact, they are also trying. Today many of the industrial houses say their attrition rate is 20 per cent to 30 per cent. So they are not able to retain the talent. So I don't know what kind of proposal they will come out with. It depends on that.

Q. And what about launch vehicles?

A. Launch vehicles, I think, by and large the final integration will be done in-house by ISRO.

Q. Always?

A. Mainly the criticality and the sensitivity of the technology is a concern. We don't want that to be floating around outside ISRO.

Q. But satellite...

A. Satellite also. See, there are certain things like gyroscope, propulsion systems – those things you know we do in our own facilities. But hardware fabrication, that the industry will do. That is why earlier I told you that 60 per cent and 40 per cent, that work content...

Q. So how much can go out? 60 per cent?

A. I would say up to 80 per cent.

Q. And for rockets?

A. That also can reach about 70–80 per cent.

Q. But there will still be a component of ISRO, always?

A. Yes. We have studied the various models. There is the European model, the NASA model and so on. But for those countries, they have a very strong aerospace industry base. They are able to pitch in. But in India we don't have such a set-up. It will take some more years to really catch up with this kind of technology.

Q. But are you encouraging ISRO scientists themselves to set up parallel industries, because that is the easiest way technology flows?

A. I don't know, after spending 20, 25 years in science, whether he will be able to be an entrepreneur. But at the same time, you know, established agencies like Hindustan Aeronautics Limited, we are encouraging them to build rockets. They are doing part of the GSLV strap-on motors already. Whoever has the capability, we are trying to encourage them to come up with solutions.

Q. What is wrong with EDUSAT utilization? Why is that lagging behind? What is happening there?

A. Well, in almost two years' time 10,000 classrooms are getting connected. So I don't know whether this is a lag or leap. It is quite an impressive number, and then again, more than 20 institutions are participating, starting from the highest level, like management institutes, some medical colleges, engineering colleges, and at the grassroots level, even teachers training, primary schools, secondary schools. So this has expanded considerably in the last two years' timeframe. But one area that needs further attention is content generation. When you have a very powerful medium like a satellite, it is not sufficient that we just show a face and give a lecture and things like that. Some innovative way of making use of it is needed. Here we have provided for multimedia facilities. With that one can do wonders, and teachers or the persons working on it, are doing well. They are making some facilities, they are holding very lively classes. And then again, when you want to target children in primary schools or secondary schools, a different approach from the conventional classroom is needed. For this, concerned state education departments have to take the initiative. I think half of the states which have been provided with these facilities are doing well. Others, again, you know it depends on their individual interests. We enable them, but unless there is a person who can drive it, sometimes it does not catch on. But there is still scope for improvement.

Q. How much life is left for the EDUSAT?

A. Another four years are left, but it will not be limited by this, since the system is in place, we will plan for replacement as and when it is needed.

Q. Where does the challenge lie? Because some are not utilizing it. Is it a problem with the satellite, or is it a problem with the ground based facility?

A. Satellite is only the medium that provides this connectivity, and by experimentation we have arrived at a configuration which is effective in terms of bandwidth and cost. With that we are really able to make an impact on the visual lessons, and also have interactive classrooms. So the capability of the system has been fully appreciated. In Karnataka there are four channels running parallel, in Tamil Nadu also I think at the moment four may expand to eight channels. That shows that people are really enthusiastic about it and are utilizing the system. It is the user community that has to really pitch in and make use of it.

Q. But is it the same problem as we had, the teething trouble when the remote sensing satellites were put up? We had a lot of data which was accumulated, the user community was not ready. Is it the same issue?

A. As I told you, delivering a class through the EDUSAT needs a certain amount of preparation. In every state they have to get equipped for that. As you know, education is a state subject. We are in dialogue with them, we are helping them; almost all the states have been provided with hubs so they can uplink the programs. And then we have provided so that major districts are connected with interactive classrooms. It can be used for educational programmes, for teachers' training and so on. Then I think Amrita Institute has come forward, they are even bringing Nobel laureates to their studio and beaming to about 50 institutions in the country. So I think it is well utilized.

Q. The user community feels otherwise.

A. Depends on who you talk to. There are some people who sleep, they will make all kinds of comments. But there are people you know…in fact, I am flooded with letters every week. So many people want to establish new connections.

Q. So you are upbeat about ISRO now?

A. Well, by God's grace, things are going all right.

Q. Will they go all right always?

A. Space is a highly risky game. But our people try to do their best and provide the best solutions to the country.

Q. So you will not fail the country?

A. We have not, so far.

Q. Can I say ISRO will not fail the country?

A. Yes, certainly.

Annexure 2
To The Moon, Together
Interview with Dr Michael Griffin, Administrator, NASA
by Pallava Bagla in Hyderabad

Q. Why go to space at all? It's a dangerous place.

A. Well, when the first people came to India, I think you could have said the same thing about India. Anytime human beings move into a new territory they have to learn new things about the new land in which they find themselves and we have done that now over most of the surface of the earth and we are beginning to expand in space.

Q. For the last few decades we have seen NASA lead what was literally the first race to the moon. In this second burst of activity happening towards the moon, it's the Asian nations, Japan, China and India, which are kind of leading from the front. Do you feel bad that NASA has been left out?

A. Well, I don't think we have been left out. We are building the Lunar Reconnaissance Orbiter, and we are embarking on a programme to take people back to the moon by the end of the next decade. I don't feel left out. I think space exploration is something that the whole human race wants to do.

Q. Do you think India is ready to go to the moon and beyond?

A. Absolutely. India has the technical capability, and will soon prove it, to send an unmanned scientific spacecraft to the moon. And frankly India has the intrinsic capability to, within quite a few years, be conducting human space flights if they wish to do so.

Q. You have also been into some of the facilities of the Indian Space Research Organisation. How do you rate them? You travel across the world, see many facilities. Are they good, bad, indifferent?

A. Well, they are good, and more important than the facilities is the quality of the engineers and scientists and the technicians. You have a great capability.

Q. Can one hope to have an Indian citizen join NASA's manned programme at some point?

A. One certainly can hope. The space station programme is nearing completion, and India did not become involved in that, but we are, as I said, beginning a new programme centred around taking people out beyond the space station, beyond earth's orbit, to the moon and then in a couple of decades, to Mars. And I will hope that the Indian people would want to be among them.

Q. You had discussions with ISRO officials even on this visit. Is there any proposal that the two countries will be working together on the manned mission?

A. Not yet. I welcome it when you decide to do it. But that is a matter for you and your people to decide.

Q. India wants to go into low earth orbits by 2014, and then probably land an Indian using an Indian rocket on the moon by 2020. Is NASA hoping to join on that programme?

A. No, I think we will not join the Indian programme, we might very well invite you to join with us and if you choose to fly on your own, we wish you well.

Q. Why don't you want to join with the Indians?

A. Well, we prefer to think about ourselves as leaders of space flight programmes, rather than followers, as I am sure you do, and we have our own programme.

Q. Well, there is at least one programme where you don't have a significant voice, where you are joining in, which is the first mission to the moon by India, Chandrayaan-1.

A. And we are very happy to be on board that, too; our science instruments were accepted in a competitive proposal and we are thrilled to be included.

Q. Are you happy getting a free ride to the moon on it?

A. Well, it is not free, but yes, we are very happy to be part of the programme.

Q. Well, you are certainly getting a free ride, you are not paying for the travel time, you are paying for the payload.

A. Not paying for the rocket, but for the instruments, yes, and of course in return we contribute data to the home nation.

Q. Are you happy with the collaboration so far on the moon mission?

A. I am delighted with it. It has been nothing but good.

Q. The scientist to scientist collaboration seemed to have worked wonders, but at the level of the government, you almost missed that ride.

A. I think there is some concern and frustration in India with our export control rules and export control laws, and I understand that frustration. It is frustrating for us as well. I would be the very first to tell you that those controls are frustrating for us. I am an engineer, I am not a politician. I have been involved at the working level in these kinds of deals and they are frustrating and we are very concerned about the proliferation of missile technologies around the world, such that other new munitions might be used by people who wish us harm or who wish to harm our allies.

Q. You think India wishes harm to America?

A. No, of course not.

Q. Last time you visited India you asked the same questions about the technology control regimes.

A. Yes.

Q. And you said you'll get the message back home.

A. And I have done that.

Q. Has it not been heard?

A. Of course it has been heard, but the decision remains within the United States that we are more concerned about controlling the proliferation of technologies for weapons of mass destruction than we are about the ability to do cooperative programmes together.

Q. There is a new tango which is happening between India and the US and the only irritant between these estranged democracies is the nuclear issue.

A. Yes, it is 25 years old, and I think we have tried to now get past that, and tried to again re-embrace closer ties. Because yes, India is the world's largest democracy and India and the United States are, we believe, natural allies and should be.

Q. Natural partners?

A. Natural partners, natural allies.

Q. You are taking it a step beyond?

A. No, I mean allies and partners to me are nearly synonyms. We are, I think, united by common goals. I use those words interchangeably.

Q. When do you hope to have man on the moon again?

A. Well, by 2020 we hope to have man on the moon again.

NASA administrator Michael Griffin, April 2007.
Photo courtesy: NASA/Renee Bouchard

Q. So India and America are racing together? India also wants to do so by 2020.

A. I have not heard that from India, but if so, I welcome you and certainly we are not racing. I hope that we can go together.

Q. Is there a real effort to work together?

A. Not yet, but there could be. We are working together on Chandrayaan, we are working together on other science programmes, and who knows what the future will bring?

Q. But beyond Chandrayaan there is no active collaboration which is concrete and planned. Chandrayaan is where everything ends, I am told. Everything else is potential.

A. Well, it's not uncommon. Let us see how this does and then we will see if we can activate some of that potential.

Annexure 3

The Moon Is Serious Business

Interview with Dr Wu Ji, Director, Centre for Space Science and Applied Research, Chinese Academy of Sciences, Beijing

Interviewed by Pallava Bagla in Udaipur, India and Beijing, China

Q. What is the genesis of the Chinese lunar programme?

A. The lunar programme has been proposed to the Chinese government since 1992 so that was the first round, but we didn't succeed at that moment. The man in space programme started from 1992. So at that moment the government thought we have to focus on one major topic and not both manned and lunar, it is too much. But at that time the scientists had already proposed to the central government that we should have a lunar programme.

Q. First came the manned programme, then the lunar programme.

A. That's the way the government decided. But parallelly we have proposed in 1992, so that was the first round, we did not succeed in the lunar programme and then the scientists didn't stop to study the lunar missions and there are many small fundings from government to support the pre-studies. So we keep going in this, pre-study upto 2001 and then there were some more formal proposals from different levels, at institutional level, from Chinese Academy of Sciences level and also in China. We have high technology programme called 863.

They are small programmes and supporting some pre-studies and high technology innovations. So this was also supported by 863 programmes. So the study became more and more clear.

So the current scientific objectives, the four scientific objectives of the Chinese lunar mission have been more or less formalized. The big breakthrough was the trans-man space flight, when the first astronauts came back by the end of 2003, then this mission became more visible. The manned mission finished the first step with these first astronauts. So the government started to think about lunar missions and by 2004, I cannot remember which month, the government formally approved the Chinese lunar programme, exploration programme and not manned lunar but exploration programme. So about the same time the three steps are formed. The first step is orbiter, to have a survey of the moon, and the second phase is to have the landing with the rover and take another five years and then it is planned to have sample return, unmanned sample return, although there are samples from the moon. But not all the places have been sampled and also we would like to demonstrate our technology to have a sample coming back from the moon. This of course will leave a very solid base for further deep space exploration.

Q. Which is the exploration orbiter and what is the budget for that?

A. Yes. For orbiter the total amount is two billion Chinese yuan. Roughly.

Q. This includes the launch or this is satellite plus launch?

A. Everything. The launch, the satellite, the payload, operation and some of the ground facilities, because we have some of the ground facilities but not all.

Q. How many launches in this two billion yuan?

A. Only one. This is the starting. There are a lot of facilities to be paved up. The satellite itself may not cost so much. I don't know the break-up but overall budget is roughly two billion Chinese yuan. So at this moment we are applying to the central government for the second phase, but it is still in the preparation phase.

Q. Which is for the rover?

A. Lander and rover. On the lander itself we have some payload, then there will be the rover coming up from the lander. There will also be some payload on the rover so both the lander and the rover are quite important to us. I believe the government may give a final green light to the second phase after our first launch succeeds. So this is a very natural mechanism. But we are preparing for the second phase and there is a lot of international interest to participate in the second phase of the trans-lunar programme.

Q. So what is the final weight of the Chang'e satellite?

A. Its weight, if I remember, is 1.9 ton or nearly 2 ton. Almost half of it is fuel.

Q. Half of it is fuel?

A. Yes. The net weight is about 1 ton.

Q. And what is the payload weight?

A. The payload weight is 130 kg.

Q. And how many payloads in all?

A. The payload has four scientific objectives. The first one: to have a complete map of the surface with three-dimensional images. We have two lenses looking forward and backward, and a stereo-camera to go as far as possible to the higher latitude. But of course on the two poles we don't have a light. So this is the first objective. Because up till now there is no complete map of the lunar surface.

And the second is to measure the material abundance of different materials, so we have three spectrometers. One operating in visible light and the second the X-ray spectrometer and the third one is a gamma-ray spectrometer. So those are three instruments for the material abundance.

Q. Does it also include Helium-3 mapping?

A. Not possible, I don't think it is possible. It is only the major materials, may be more than ten of the different types of material.

Q. Fourteen of them!

A. More than ten, but in the beginning they say we can measure more than 15 or 14 but some of them are really difficult.

Q. So nobody has done this kind of mapping of ten different materials?

A. People are more interested in five of the materials. So we would like to increase the number finally to ten, to map different materials, distribution of different materials. The third objective is to evaluate the thickness of the lunar soil by four frequency band radiometer, microwave radiometer from 3 GHz upto 37.

Q. The mission has a life of one year.

A. This mission has a life of one year.

Q. Any major challenges you have to overcome in the payload preparations for Chang'e?

A. The temperature is really a big challenge because there is no atmosphere and the reflection from the moon is very strong, so the part of the spacecraft facing the moon is getting hot and the other face when it is not illuminated by the sun is getting very cold. So in the orbit if we don't have the sun, the temperature will come down very much.

Q. Is there any international payload on Chang'e-I?

Dr Wu Ji, Executive Director, Center for Space Science and Applied Research, China, with a lunar globe in his hand, Udaipur, 2004

A. No, no. As I said before, it was a quick decision. We would like to have only Chinese payloads.

Q. How many missions are planned for Chang'e if you get the funds? One, two, three...

A. In the second phase, I don't know, I think that there should be at least two missions. First one lander and second one, lander with a rover.

Q. Meaning you have Chang'e orbiter one and then you have the lander and rover.

A. Some would return but in a row. But I don't know how it'll be called, may be Chang'e-2 or Chang'e-3. I don't know. Orbiter at this moment we have only one but we have a back-up. So if we fail with this launch we may have the second launch. And for the second phase, we have at least two, maybe three, depends on the funding.

Q. And sample return? How many do you want to do?

A. It is a little bit too far, I don't know.

Q. And what about a man on the moon, how many missions have you planned for the manned mission to the moon?

A. For the manned mission a lot of scientists are interested, but the government doesn't really think about that. It still remains a scientist's proposal. Not even proposal, lot of scientists talk with television people, we will like to have manned mission. Maybe in the future the Chinese manned programme and lunar exploration programme can merge into one direction. At this moment they are two parallel programmes. In the manned mission they have their free walk in space, rendezvous and docking and in lunar mission we have lander and rover. But at this moment they are going parallel.

Q. Walk, and what did you say is the second part?

A. Rendezvous and docking. Of course, with docking you need a space lab.

Q. Why are they so independent, the moon mission and the manned mission?

A. It is natural as they start from two different points. They have their own proposals, they have to plan their future, so they have planned their future separately. But maybe at a certain point they may merge. At this moment they are held parallelly. It is quite natural to have these two goals separate. Because when the man in space mission started nobody talked about the moon, the government didn't think about it. So they have their own roadmap already and we have the lunar mission. The manned mission had not started and had not succeeded yet. So in order to avoid any influence between them, they have their own roadmap. These two are quite naturally planned in this way. More and more people talk about merging but at this moment nobody, at the government level, there is no proposal even to do this because both of them have a lot of work to do. They should do their job and show good results.

Q. What would be the role of NASA or the US in these, both the manned and the lunar programmes? Is there a role being defined because Michael Griffin of NASA came to China and is there more collaboration between China and the US?

A. Like I said, Michael Griffin has visited us, it is just a starting point, it has not been like this for many years. I do not know how many years, more than 15 years, something like that. And he visited us and showed a very active attitude to collaborate with us and from the Chinese side, of course, we are also very keen to collaborate with the US but it is better to start with some more neutral plans, for example scientific programmes. But not first step to start with a manned programme. A manned programme requires lot of transparency and costs and after so many years of suspending our relationship with NASA we are just coming back. So it is better to have something to start with that is more solid, have more trust and be more neutral. So I think the space science mission is one of the best fields to start with. Personnel exchange, data exchange and put our payload on their mission and be prepared to carry theirs on our mission – from this level.

Q. Is this an exciting time in China for space?

A. Not so much. We still need more funding from the government. So if the government can foot more funding for science we are really growing, but as far as the growing, at this moment it is not so big. So if we have more money we will have more opportunities to collaborate with international partners.

Q. How much more money do you need?

A. What we spent is even less than ISRO. So you can imagine. We have had Double Star satellite, and after Double Star, we have no money to have the second space physics mission and the lunar is taking a lot of money. In the next five years, we will have only one or maybe two space science missions. One is the astronomy telescope, X-ray telescope, which is more or less fixed and the second one is macro-gravity retrievable satellite. That is in the next five years. There might be only two scientific missions apart from the manned mission and the lunar mission. And in the previous five years we had Double Star and we started the lunar mission and a lot of experiments with the manned mission.

Q. For the manned mission is it an exciting time?

A. The manned mission is going on and, as I have said, they have their programmes. So next is the free walk or Ex-Vehicle Activity.

Q. So is that exciting? There was a lot of excitement in the world?

A. I think it is an exciting activity. We need to develop our own space suit for EVA and a lot of new activities and attract the public attention to this manned mission so there would be another big time, I think, after first astronauts.

Q. What is motivating China for these space missions?

A. The government is thinking space science is something good to boom up the economy and to have young people come to science, to learn science and engineering and to have spinoffs in other industries. I think they understand this. And we Chinese as a contribution to the world history – we have many inventions like gunpowder, remember this – and we should contribute more as a major nation. India is also a major nation and if you look at space, the history of space exploration, the US and Russia almost taking 99 per cent of the first exploration, first satellite, first manned space, first lunar exploration, first landing on Venus, first landing on Mars. Almost 99 per cent comes from the US and very little from Europe even. So I think China and India should make more contributions.

Q. India's mission is for putting a man in orbit. They are saying 2014, they want to put a man in space.

A. That is great! That is surprising. This will be fantastic, even before the US. The US is coming back to the moon in 2018, and the Japanese announced 2020, and Europeans announced 2025. So if India can go to the moon by 2014, that is fantastic.

Q. 2014, man in space, and 2020, man on the moon.

A. Even 2020 is fantastic. Congratulations to your government. This is a very broad, ambitious programme. This will definitely inject a lot of energy into international space activities. It will encourage a lot of governments, maybe even the Chinese government, to think of more space activities.

Q. Can the two countries afford this space programme? Both India and China are very poor.

A. Yes, your economy is going up and also currently you are spending quite a lot of money on space as compared to other developing countries. I think on civil space you are spending even more than China and also the space industry in India is concentrated. It is not the market economy, you are planning everything under ISRO. It is more planned economy and the developing countries like us, it might be a good idea to concentrate all the resources with one agency.

In fact, in China we are starting market economy even in our space sector a little bit now, because more and more industries want to come in, like in the US there are many space industries, they compete with each other. But in India if you spend your limited resources on one agency, you might do things much faster than the US and more efficiently. So I think you don't have to spend an enormous amount of money but increase your budget step by step. But of course other missions may suffer a little bit, it might reduce the spendings on other missions.

Q. Is there competition between India and China? India announced the moon mission, China put a manned mission and India has decided to have a manned mission; it almost seems like a real race to space. What is your opinion?

A. It is not like a cold war because relations between us are very good and are getting better and better. The relations between China and the US are also getting better and we have traditionally good relations with Europe and the Russians. So any country has their activity, I think it injects energy into this international community. Top leaders should think about how useful those space activities are for their own nation, for the education of the young generation and to bring high technology development spinoffs of the space industry to other industries and I think the US Apollo programme has spent 'one dollar gain fourteen dollars' as they said.

It is a very useful tool, the space programme, to bring the nation up. This, the top leaders say, is the way. So it is not a competition. They may not want to lose the chance or if they use this activity to trigger the national economy, we should do that too, otherwise we lose the chance. So in this way it helps each other to move up. This is my opinion.

Q. Is there any likely collaboration with India?

A. We certainly wish so, we would like to have a collaboration with India. We should find more opportunities to talk with the Indian space agency, ISRO, and we should know more about India's space science programmes, like what we did last time at international meetings in India. A lot of Indian scientists would like ISRO to have a space physics mission, not only astronomy, so suddenly if you have another space science mission we would like to propose to have our own payload or any kind of cooperation with Indian space programmes. If India has another science mission, for example space physics or solar physics or any other space exploration, we would like to participate at your invitation. Of course we have between scientists a lot of collaboration, lots of good relations and ground-based observation. Recently, we had the international helio-physics coordinating meeting in China and all the ground-based observation people, for example, Japanese and also Indian, come here and discuss and make coordinated ground-based measurement of space environment on solar observations and so on together to share the data. So these are going very well.

Q. You don't see India as a competitor?

A. No, why?

Q. Everybody thinks India and China are competitors.

A. No, I think maybe in economic area but in science, of course, the more people do it, the better, and if you do this, we would do other things and we share the data and we cover more areas. So in science collaboration is always wanted.

Q. Can you hope for a joint mission between India and China?

A. Okay, I would love to have that alternative, so between agencies they should talk more and study.

Endnotes

Introduction

1. Bagla, P., The *Indian Express*, 12 May 1999.

Chapter 1 – Reaching for the Unknown: The Moon's Origins and the Search for Water

1. *Chandrayaan-1: The Indian Mission to the Moon: Project Report*, ISRO, August 2003.
2. Datta, J. and S.C. Chakravarty, *Chandrayaan-1: India's First Scientific Mission to the Moon*, ISRO, July 2007.
3. Agrawal, P.C., B.V. Sreekantan and N. Bhandari, 'Space Astronomy and Interplanetary Exploration', *Current Science*, Vol. 93, No. 12, 25 December 2007 (Special Section: Indian Space Programme).
4. *The Scientific Context for Exploration of the Moon – Interim Report*, National Research Council of the National Academies, Washington DC: The National Academies Press, 2006.
5. Taylor, S.R., 'Lunar Science: An Overview', *Journal of Earth System Science* 114, No. 6, December 2005.
6. Lawler, A., 'Moon Maintains Its Mysteries', *Science*, Vol. 300, 2 May 2003.

7. Agrawal P.C., B.V. Sreekantan and N. Bhandari, 'Space Astronomy and Interplanetary Exploration', *Current Science*, Vol. 93, No. 12, 25 December 2007 (Special Section: Indian Space Programme).
8. *The Scientific Context for Exploration of the Moon – Interim Report,* National Research Council of the National Academies, Washington DC: The National Academies Press, 2006.
9. www.isro.gov.in
10. *The Scientific Context for Exploration of the Moon – Interim Report,* National Research Council of the National Academies, Washington DC: The National Academies Press, 2006.

Chapter 3 – The World's Many Missions

1. www.esa.int
2. Matogawa, Y., 'Looking Ahead: Half a Century of Space Activities in Japan'. In: Rao, P.V. Manoranjan (ed.) *50 Years of Space: A Global Perspective,* Astronautical Society of India, Universities Press (India) Pvt Ltd, 2007.
3. www.jaxa.jp
4. Normile, D. and P. Bagla, 'Asian Powers Shoot for the Moon, with Orbiting Research Missions', *Science, 31* August 2007.
5. Lawler, A. and P. Bagla, 'Asking for the Moon', *Science,* Vol. 315, 16 March 2007.
6. www.cnsa.gov.cn
7. Normile, D. and P. Bagla, 'Asian Powers Shoot for the Moon, With Orbiting Research Missions', *Science,* 31 August 2007
8. Ge, L. and Z. Yao, 'Following an Indigenous Path: China's Space Activities from 1956 to 2006'. In: Rao, P.V. Manoranjan (ed.) *50 Years of Space: A Global Perspective,* Astronautical Society of India, Universities Press (India) Pvt Ltd, 2007.
9. *Chandrayaan-1: The Indian Mission to the Moon: Project Report,* ISRO, August 2003.

10. Lawler, A. and P. Bagla, 'The New Race to the Moon', *Science*, Vol. 300, 2 May, 2003.
11. Schmitt, H., 'Return to the Moon', *Exploration, Enterprise and Energy in the Human Settlement of Space*, Copernicus Books in association with Praxis Publishing Ltd., 2006.
12. Lowman, Paul D. Jr., 'Our First Lunar Program: What Did We Get From Apollo?', www.nasa.gov., downloaded on 6 September 2007.
13. www.nasa.gov

Chapter 4 – Beyond the Stratosphere: India's Space Programme

1. Medvedchikov, A.I., 'Pioneering Space: The Russian Space Programme'. In: Rao, P.V. Manoranjan (ed.) *50 Years of Space: A Global Perspective*. Astronautical Society of India, Universities Press (India) Pvt Ltd., 2007.
2. Ibid.
3. www.dst.gov.in
4. www.isro.gov.in
5. Rao, P.V. Manoranjan, 'No Ambiguity of Purpose: The Indian Space Programme.' In: *50 Years of Space: A Global Perspective*.
6. Suresh, B.N. 2007. In: *50 Years of Space: A Global Perspective*.
7. Rao, P.V. Manoranjan. 2007. 'No Ambiguity of Purpose: The Indian Space Programme.' In: Rao, P.V. Manoranjan (ed.) *50 Years of Space: A Global Perspective*. Astronautical Society of India, Universities Press (India) Pvt Ltd., 2007.
8. Nair, G. Madhavan. In: Foreword to Rao, P.V. Manoranjan (ed.) *50 Years of Space: A Global Perspective*. Astronautical Society of India, Universities Press (India) Pvt Ltd., 2007.
9. www.antrix.gov.in
10. Kasturirangan, K., 'Dr Vikram Sarabhai: The Light that Leads the Indian Space Programme', *Resonance*, December 2001.

11. Gandhi, D., 'Is Edusat a Turkey in the Sky?' The *Hindu*, 13 March 2007.
12. Bagla, P., 'Remote Sensing: India Struggles to Handle an Embarrassment of Riches', *Science*, Vol. 273, 26 July 1996.
13. Sarabhai, V.A., 'Sources of Man's Knowledge', National Programme of Talks Series, Bangalore, *Resonance*, December 2001.

Chapter 5 – Chandrayaan: India's Shot at the Moon

1. Datta, J. and S.C. Chakravarty, *Chandrayaan-1: India's First Scientific Mission to the Moon,* ISRO, July 2007.
2. Bagla, P., 'Destination Moon – India's Next Big Thing', The *Indian Express*, 12 May 1999.
3. Bagla, P., 'India Okays Moon Mission', *Science*NOW, 19 August 2003.
4. Bagla, P., 'Destination Moon – India's Next Big Thing', The *Indian Express*, 12 May 1999.
5. Datta, J. and S.C. Chakravarty, *Chandrayaan-1: India's First Scientific Mission to the Moon,* ISRO, July 2007.
6. Bagla, P., 'Indian Agency Eyes Moon in Shift Towards Research', *Science*, Vol. 286, 17 December 1999.
7. www.isro.gov.in
8. Chengappa, R., 'Moon Mission', *India Today*, 3 July 2000.
9. Datta, J. and S.C. Chakravarty, *Chandrayaan-1: India's First Scientific Mission to the Moon,* ISRO, July 2007.
10. Ibid.
11. Ibid.
12. Ibid.
13. Narlikar, J.V., 'Asking for the Moon: Back to Basics in Space', The *Times of India*, 30 May 2003.
14. www.isro.gov.in
15. Bagla, P., 'Indian Lunar Probe Goes Global', *Science*, Vol. 303, 27 February 2004.

16. Bhandari, N., (ed.) 'Proceedings of the International Conference on Exploration and Utilisation of the Moon (November 22-26, 2004)', *Journal of Earth System Science*, Vol. 114, No. 6, December 2005.
17. Ibid.
18. Thyagarajan, K. and M. Annadurai, 'Chandrayaan-1. The First Indian Mission to the Moon', *Journal of Aerospace Sciences and Technologies*, Vol. 58, No. 4, November 2006.

Chapter 6 – The Future Is Here

1. Bagla, P., NDTV, February 2007.
2. www.pib.nic.in
3. Bagla, P., NDTV, February 2007.
4 Bagla, P., 'Moonstruck', *Science*, Vol. 318, 16 November 2007.
5. www.isro.gov
6. Subramanian, T.S., 'Mission Mode', *Frontline*, 18 January 2007.
7. http://sci.esa.int/ilewg/
8. Schmitt, H., *Return to the Moon: Exploration, Enterprise and Energy in the Human Settlement of Space*. Praxis Publishing Ltd., 2006.
9. Ibid.
10. www.nasa.gov
11. Datta, J. and S.C. Chakravarty, *Chandrayaan-1: India's First Scientific Mission to the Moon*, ISRO, July 2007.
12. www.nasa.gov (Official press release: NASA Tests Lunar Habitat in Extreme Antarctic Environment, November 2007).
13. Lawler, A. and P. Bagla, 'Asking for the Moon', *Science*, Vol. 315, 16 March 2007.
14. http//www.nasa.gov/mission pages/constellation/altair/index.html
15. Sarabhai, V.A., 'Sources of Man's Knowledge', National Programme of Talks Series, Bangalore, *Resonance*, December 2001.

16. Sridhara Murthi, K.R., et al, 'Legal Environment for Space Activities', *Current Science*, Vol. 93, No. 12, 25 December 2007 (Special Section: Indian Space Programme).

Epilogue

1. Bagla, P., 'Pileup on the Moon', *Science*, Vol. 318, 12 October 2007.
2. Ibid.

Bibliography

1. *Chandrayaan-1: The Indian Mission to the Moon: Project Report*, ISRO, August 2003.
2 Datta, J. and S.C. Chakravarty, *Chandrayaan-1: India's First Scientific Mission to the Moon*, ISRO, July 2007.
3. Agrawal, P.C., B.V. Sreekantan and N. Bhandari, 'Space Astronomy and Interplanetary Exploration', *Current Science*, Vol. 93, No. 12, 25 December 2007 (Special Section: Indian Space Programme).
4. *The Scientific Context for Exploration of the Moon – Interim Report*, National Research Council of the National Academies, Washington DC: The National Academies Press, 2006.
5. Rao, P.V. Manoranjan. 2007. 'No Ambiguity of Purpose: The Indian Space Programme.' In: Rao, P.V. Manoranjan (ed.) *50 Years of Space: A Global Perspective*. Astronautical Society of India, Universities Press (India) Pvt Ltd., 2007.
6. Sarabhai, V.A., 'Sources of Man's Knowledge', National Programme of Talks Series, Bangalore, *Resonance*, December 2001.
7. Bhandari, N., (ed.) 'Proceedings of the International Conference on Exploration and Utilisation of the Moon (November 22-26, 2004)', *Journal of Earth System Science*, Vol. 114, No. 6, December 2005.
8. Thyagarajan, K. and M. Annadurai, 'Chandrayaan-1. The First Indian Mission to the Moon', *Journal of Aerospace Sciences and Technologies*, Vol. 58, No. 4, November 2006.

9. Schmitt, H., *Return to the Moon: Exploration, Enterprise and Energy in the Human Settlement of Space.* Praxis Publishing Ltd., 2006.
10. Sridhara Murthi, K.R., et al, 'Legal Environment for Space Activities', *Current Science*, Vol. 93, No. 12, 25 December 2007 (Special Section: Indian Space Programme).

WEBSITES :

1. www.isro.gov.in
2. www.antrix.gov.in
3. www.cnsa.gov.cn
4. www.esa.int
6. www.jaxa.jp
7. www.nasa.gov
8. www.sci.esa.int/ilewg
9. www.pib.nic.in
10. en.wikipedia.org/wiki/Hindu_calendar
11. www.webonautics.com/mythology/soma.html
12. www.sanskrit.org/www/Astronomy/Rahu.html
13. www.indiasite.com/festivals/
14. lunar.ksc.nasa.gov/history/mythologyh.html
15. *Windows to the Universe* at www.ucar.edu/

Index

Acknowledgments

This book celebrates the captivating, thrilling world that is space science and exploration. The journey to get to the point in one's knowledge and understanding to be able to document, unravel and tell the story of this complex world and its even more complex missions and projects has been a long one. Needless to say, there are many individuals who have made crucial contributions during the journey, and this is only a modest effort to thank them.

Over the years, reporting from deep within ISRO would not have been possible without the able support and guidance of Dr K. Kasturirangan who conceived India's Chandrayaan mission and is now an able parliamentarian. He always provided a perspective when approached, readily and with amazing grace. We acknowledge Dr G. Madhavan Nair who gave ISRO its second vision, for offering unprecedented access. And S. Satish, ISRO's spokesperson for the press, who is always just a phone call away and ever willing to help.

In the past two defining decades at ISRO, Dr S. Krishnamurthy, the media interface for ISRO, was a constant sounding board on diverse issues, and always around to aid the critical process of 'telling the story' as it was. Before he died in a road accident en route to a launch, Kitta was the public face of ISRO and we would like to place on record our deep appreciation of his support over the years. Also, the late Rajeev Lochan, whose command over international space affairs is missed. Lochan, too, lost his life in that accident.

Special thanks to Dr B.R. Guruprasad for sharing his encyclopaedic knowledge of Indian and foreign missions. We would like to acknowledge Dr Narendra Bhandari and Dr George Joseph, both early stalwarts of the Indian space programme and experts who have freely given of their time in supporting the completion of this book. We also acknowledge Prof. U.R. Rao, Dr B.N. Suresh, K.S. Sridhara Murthi, Dr K. Radhakrishnan, Dr V.S. Balagangadharan, Abhay Kumar, Bhaskar Narayana, Jayaraman, Radhika Ramachandran, S.V. Kibe, V. Sunderaramaiah, V. Adimurthy, A.K. Shivkumar, Dr J.N. Goswami, S.C. Chakravarty, Jayati Datta, Madan Lal and Yashoda. Without their support over the years, this book would not have been possible.

We record our appreciation for Dr A.P.J. Abdul Kalam, who has always supported coverage of science and technology, and in particular the space and atomic energy programmes. Special thanks to Prime Minister Manmohan Singh, who has readily answered tough questions on the Indian space and atomic programmes. To Union Minister for Science and Technology Kapil Sibal, for the freshness of approach he brings to his job and his consistent encouragement of good writing in the press, and to Prof. Murli Manohar Joshi, former union minister for science and technology. Also to Jairam Ramesh, engineer-turned-politician, whose skill in explaining complexities is unsurpassed.

Thanks to the two accessible senior technocrats of the Indian Science and Technology (S&T) establishment – Dr V.S. Ramamurthy and Dr Raghunath Mashelkar – for bringing passion to the management of S&T. If it were not for them, reporting on S&T would be a very different story from what it is today in India. Professors Yashpal and M.G.K. Menon, pioneers of the Indian space movement, were always generous with their time and their knowledge. Also thanks to Dr Anil Kakodkar, Dr P.S. Goel, Dr S. Banerjee, Dr T. Ramasami, Dr Samir Brahmachari, Dr M.K. Bhan, Prof. C.N.R. Rao, Dr R. Chidambaram – individuals who are leading from the front and have always been accessible, keeping their doors open for communication. We would also like to thank Dr D. Balasubramanian, Dr S.K. Malhotra, Dr Aiyagiri Rao and Dr S.K. Sikka.

Thank you to Dr Wu Ji and Dr Michael Griffin for being so generous with their time, and for providing detailed inputs to help us achieve a better understanding of the programmes in China and the US respectively, as also a better global perspective.

A good science writer always seeks credible information, which in turn is deeply influenced by the scientists he or she interacts with. Some invaluable interactions made this work on lunar exploration and Chandrayaan possible, for which we would like to thank Dr Jayant V. Narlikar, Dr Ramnath Cowsik, Prof. Govind Swarup, Dr Paul Spudis, Dr Carle Pieters, Dr Stas Barabash, Dr Manuel Grande, Dr Anil Kulkarni, Dr Bernard Foing, Dr Urs Mall and Dr Tsvetan Dachev.

In the journalistic world, no amount of reportage is worthwhile without the support of a receptive news organization with sensitive editorial leadership that understands the real value – and the broader context – of a story. A cornerstone of this effort has been the support from 1994 onwards, of bright and demanding colleagues at *Science* magazine. Special thanks to Colin Norman for his trust, Elliot Marshal for his unstinting support and Jeffrey Mervis for shaping early reportage and giving it the right appeal and very importantly, Richard Stone, first Asia editor of *Science* based in Beijing, who seeks out wider and better coverage from the two emerging giants of the world, China and India. Also thanks to other colleagues at *Science*: Andrew Lawler (whose insights into NASA have been very helpful), Dennis Normile, Constance Holden, Eli Kintisch, David Malakoff, Ginger Pinholster and Yudhijit Bhattacharya.

Special thanks to NDTV's Dr Prannoy Roy and Radhika Roy for unfolding a totally new facet of science in television journalism in India. To Barkha Dutt and Sonia Singh for ensuring that only the highest quality of reportage gets aired and for pushing the frontiers of public understanding of science. To Manika Raikwar and Chetan Bhattacharjee for always being there when guidance was required, and to Nitin Gokhale, Amitabh Revi, Vishnu Som and Ajai Shukla for their understanding of complex technical issues, and for so effectively communicating them to the viewer. Thanks, too, to Arjun Hardas at NDTV Profit, whose insight into space issues is admirable. At NDTV India, thanks to Manish Kumar, Sanjay Ahirwal, Ravish Kumar and Sunil Saini for ensuring that the Hindi audience does not miss out.

At the *Indian Express*, thanks to Shekhar Gupta for having provided a platform. To Raj Kamal Jha, executive editor of the paper, a very special appreciation for his sharp eye, and for recognizing the value of good science and technology coverage. Thanks also to Unni Rajan Shankar for his support. And to colleagues at the *Science Reporter* in the early days – Bal Phondke, Biman Basu, the late Dilip Salwi, Hasan Jawaid Khan, Kollegala Sharma and V.S. Chaturvedi.

Thanks to colleagues from the Indian journalistic community – especially T.S. Subramanian and Srinivas Laxman – for tracking space developments with such rigour. Also to Sondeep Shankar, Vinod Sharma, Mahesh Rangarajan, Vijay Naik and Sunil Gatade, with whom one can share thoughts and brainstorm.

We would like to thank V.K. Karthika and Shivmeet Deol at HarperCollins for their hard work on the manuscript and for shaping the book into what it is today. Also thanks to Tara Goswami for a beautiful cover. Thanks to Surit Mitra, book lover par excellence, who is a great advisor and guide in the maze of the book world. Thanks also to Shammy Baweja for always having said that it is the first few lines that take the longest, the rest of the story then sings across the pages.

A virtual thanks to the Wikipedia project – a global initiative that has truly brought the library home, given research a 'mouse' edge, and provides credibility in the otherwise crazy wilderness of the net. Thanks also to Sher Shah Bagla for being the constant and vigilant companion as we burnt the midnight oil.

If it wasn't for some unshakeable and unstinting support at home, especially from our two delightful children, Nayantara and Ashwat, this book would still be on the laptop. Thanks to our mothers Sharad Bagla and Padma Menon, without whose insightful upbringing we may have been on very different paths. And to the family – Gunjan, Venu, Niven, Sunanda, Arun and Anil – and diehard friends Harsh and Prerna, Shantnu and Vibha, Amita and Bharath, and Saguna, for always being there.

All the support notwithstanding, any errors in this book, factual or otherwise, are the sole responsibility of the authors.